As - g - IV - 5 - 20

The Philippines

An Economic and Social Geography

The Philippines

An Economic and Social Geography

T.M. BURLEY

LONDON
G. Bell and Sons Ltd.
1973

Copyright © 1973
by G. Bell & Sons Ltd.,
York House, Portugal Street,
London W.C.2.

All Rights Reserved.
No part of this publication
may be reproduced, stored in a retrieval
system, or transmitted, in any form or by any
means, electronic, mechanical, photocopying,
recording or otherwise, without the prior permission
of G. Bell and Sons Ltd.

ISBN 0 7135 1706 9

Set in IBM Aldine Roman by Print Origination,
Bootle, Lancs.
Printed in Great Britain by
Redwood Press Limited, Trowbridge, Wiltshire

Contents

Author's Note	ix
Acknowledgements	xi

1 INTRODUCTION — 1
The Philippines and South-East Asia — 1
Data Presentation — 5

2 THE URBAN SCENE — 9
Manila—the National Centre — 13
Hilongos—a Minor Provincial Centre — 21
Cebu—an Inter-regional Centre — 28
Dumaguete—a Major Provincial Centre — 31
Trinidad—a Barrio — 34
Iligan—an Industrial Centre — 38

3 THE RURAL SCENE — 42
Geographical Patterns of Land Use — 42
Rice — 48
Maize — 54
Coconuts — 58
Sugar Cane — 65
Abacá — 71
Tobacco — 73
Fishing — 80
Forestry — 84
Mining — 89

4	**THE PLACE**	95
	Landforms	95
	Climate	104
	Soils	112
	Vegetation	117
5	**THE PEOPLE**	124
	Basic Features	125
	Population	138
	Land Tenure	153
	Social Attitudes	158
6	**THE REGIONS**	167
	The Cagayan Valley	169
	The Ilocos	174
	Central Luzon	177
	Manila	181
	South-Western Luzon	186
	Bondoc and Bicol	190
	The Eastern Visayas	193
	The Central Visayas	198
	The Western Visayas	202
	Eastern Mindanao	206
	Central Mindanao	212
	The Misamis Coastal Belt	218
	South-Western Mindanao	221
	Palawan and Mindoro	225
7	**THE PAST**	230
	The Pre-Spanish Era	230
	The Spanish Era	234
	The American Era	241
8	**THE PRESENT**	248
	A. Economic Geography	248
	Primary Production	251
	Manufacturing	261
	Transport and Communications	272
	Commerce	277

	B.	Social Geography	291
		The Rural Scene	294
		The Urban Scene	304
		The Political Scene	310
9	THE FUTURE		316
	A.	Economic Geography	318
		Primary Production	322
		Manufacturing	330
		Transport and Communications	336
		Commerce	339
	B.	Social Geography	345
		The Rural Scene	346
		The Urban Scene	352
		The Political Scene	357
10	CONCLUSION		360

Glossary	365
Select Bibliography	368
Index	371

Author's Note

Various important developments occurring subsequent to completion of the manuscript perhaps should be brought to the attention of the reader.

In the political field, the proclamation of Martial Law by President Marcos in late September, 1972 and the subsequent announcement in early November of a referendum to be held in January 1973 to adopt a constitution that envisages a switch to the parliamentary system of government, could in retrospect be seen as the dawning of social maturity for the nation (as discussed in Chapter 10). But, freedom from corruption, lawlessness and social injustice only will materialise if words are translated into deeds (and recent Philippine political history offers little encouragement on this score). Thus, for example, the designation of the entire Republic as a land reform area will have no immediate impact upon the social geography of the country (but see p. 320). The President has denied that his actions are unconstitutional but there can be little doubt that the democratic process has been severely jolted: nevertheless, his unprecedented measures (but not entirely unforseen: see pp. 316, 348, and 360) may well provide the incentive for a transformation of Philippine society. But, because of the deep seated nature of the problem (see p. 360), success may elude the President even though his measures have been welcomed by the man in the street, and by the middle class in particular.

The President's measures probably were precipitated by consideration of the current and future repercussions of the most disastrous series of typhoons to hit the Philippines in the past century. It will be many months before the full cost of the exceptional weather is known but, in their passage across Luzon in July 1972, the typhoons are feared to have caused more material damage to the Republic than was suffered throughout the Second World War. By early August the rains associated with the disturbances finally abated leaving 355 known dead and over 1.5 million persons homeless. In Manila the July rainfall totalled over 200 cm (80 inches) but double this amount fell over parts of the Central Plain of Luzon. Total damage to property is expected to exceed £125 million with transport and irrigation facilities, fishpond farming and, above all, rice production seriously affected. The loss of food supplies, with resultant higher prices and costly imports, will have serious short term repercussions for the Philippine economy which was showing signs of recovery following the devaluation of the *peso*.

N.B. Other recent developments have been ignored as it is considered that they would cause unnecessary confusion to the reader. Thus, in converting monetary values the *de facto* devaluation of the *peso* that took place early in 1970 and subsequent changes in exchange rates have not been taken into account. An exchange rate of £1 sterling P9.34 has been employed throughout. Also, the creation of additional provinces by the government has been ignored because little data concerning the new units has been available. This does not apply to provinces created prior to 1960, such as Lanao del Norte and del Sur, but does apply, for example, to the more recent subdivision of Mountain Province into the provinces of Benguet, Bontoc, Ifugao and Kalinga-Apayao.

London, November, 1972   T.M.B.

Acknowledgements

This book would not have been possible without the hospitality and assistance provided by Filipinos from all walks of life during my visits in 1963, 1966-7 and 1971. Outstanding in this respect are my wife's relatives in Hilongos and elsewhere in Leyte, in Cebu, in Iloilo, and in Manila. Special mention also needs to be made of E. Conrad Geeslin, the Manila representative of the Economist Intelligence Unit (E.I.U.), who went out of his way to keep me informed of latest developments.

To the E.I.U. in general and to Mr. H. Gearson, a Board Director, in particular, I owe a considerable debt for allowing me use of the company's library facilities and expertise in economic development studies. Assistance from the academic world has been substantial, in particular from Professors Cutshall, Huke, Schul, Spencer, Vandermeer and Wernstedt.

Last, but by no means least, I gratefully acknowledge the debt I owe to my wife. She has served in many capacities: guide, translator, typist, editor and proof reader to name but a few. Most important, however, has been her constant encouragement which has enabled me to surmount the various crises through which the manuscript passed.

On my shoulders alone, however, rests responsibility for the book's contents. Most Filipinos and friends of the Philippines, I trust, will find it provides food for thought. Thus, perhaps indirectly my efforts may stimulate some of the anticipated changes in the economic and social geography of the Philippines alluded to in the text.

Acknowledgements for Maps

Fig. 2 is based on E.L. Ullman, 'Trade Centers and Tributary Areas of the Philippines', *Geographical Review*, Vol. 50 (1960), pp. 203-18

Fig. 3 is from T.G. McGee, *The Southeast Asian City*, Bell, London, 1967

Fig. 7 is from A. Cutshall, 'Dumaguete: An Urban Study of a Philippine Community', *Philippine Geographical Journal*, Vol. V nos. 1 and 2, Jan-June 1957

Fig. 8 is from H. Umehara, 'Socio-Economic Structure of The Rural Philippines, *The Developing Economies*, Vol. 7 (1969), pp. 316-31

Figs. 10, 12, 13, 15, 19, 21, 22 and 36 are from R.E. Huke, *Shadows on the Land: An Economic Geography of the Philippines*, Bookmark Inc., Manila, 1963

Fig. 11 is based on Bureau of Agricultural Economics and International Rice Research Institute data

Fig. 14 is based on Philippine Sugar Institute data

Fig. 17 is based on Philippine Fisheries Commission data

Fig. 18 is from the Bureau of Forestry

Fig. 23 is based on A. Barrera, 'Classification and Utilization of Some Philippine Soils', *Journal of Tropical Geography*, Vol. 18 (August 1964)

Figs. 25 and 26 are from F.L. Wernstedt and J.E. Spencer, *The Philippine Island World: A Physical, Cultural and Regional Geography*, University of California Press, Berkeley and Los Angeles, 1967. Reprinted by permission of The Regents of The University of California

Figs. 27, 28 and 29 are based on Bureau of Census and Statistics data

Fig. 31 is from T.W. Luna, Jnr., 'Manufacturing in Greater Manila', *Philippine Geographical Journal*, Vol. VIII nos. 3 and 4, July-Dec 1964

Fig. 34 is based on C. Vandermeer, 'Population Patterns on The Island of Cebu, The Philippines: 1500 to 1900', *Annals of The Association of American Geographers*, Vol. 57 no. 2, June 1967

Fig. 35 is based on Bureau of Agricultural Economics data

Fig. 37 is from *The Fookien Times Yearbook*, 1969; statistics for 1969–71 are based on slightly different Central Bank of The Philippines series

Fig. 38 is from The Philippine Automotive Association

Fig. 42 is from D.J. Dwyer, 'Irrigation and Land Problems in The Central Plain of Luzon', *Geography*, Vol. 49 (1964), pp. 238-46

Fig. 47 is based on University of The Philippines projections

Jacket illustration by courtesy of the Embassy of the Philippines

1
Introduction

The thirty-seven million inhabitants of the Republic of the Philippines have at their disposal some 1·25 million square kilometres of the Western Pacific containing a land area of 300 000 square kilometres. In size therefore the Philippines is equivalent in area to Italy or Arizona or, where comparable island nations are concerned, is slightly larger than New Zealand but slightly smaller than Japan. The largest island, Luzon, is comparable in area to Guatemala or Bulgaria; totalling 107 000 square kilometres it contains approximately 35% of the land surface of the Philippines (Mindanao, with 97 000 square kilometres, accounts for a further 32%). The Republic is made up of more than seven thousand islands, some eight hundred of which are inhabited. This archipelago is primarily north-south oriented and extends through sixteen degrees of latitude (Fig. 1). The islanders of the Sulu Sea in the extreme south are thus placed some 1750 kilometres distant from their compatriots who live close to Taiwan in the Batanes Islands.

The Philippines and South-East Asia

Geographically, the Philippines is a part of South-East Asia although it is situated on the eastern margin of this vast area of land and water. The war in Vietnam has placed this part of the world very much in the public eye in recent years but the area is worthy of careful attention for other reasons. Politically, South-East Asia can be compared to the Balkans in the early twentieth century; commercially, it epitomises the upheaval and subsequent reorientation of business and trading relationships consequent to the withdrawal of the colonial powers; and, in social terms, national independence has accelerated the evolu-

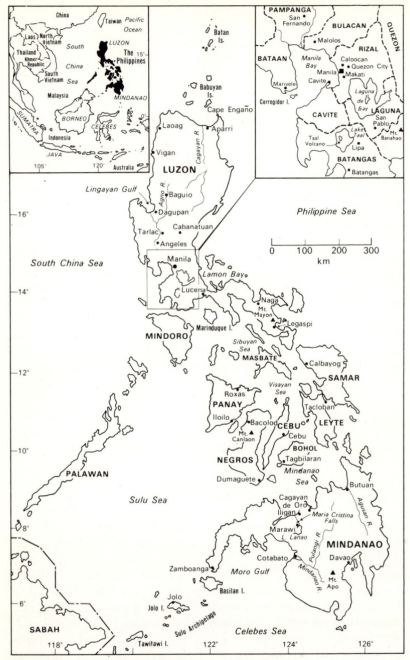

Fig. 1. The Republic of the Philippines

tion of its own institutions which are the product of a unique blending of Eastern and Western cultures. Natural conditions, too, are both impressive and spectacular, the area's volcanoes, monsoon climate, equatorial flora and fauna, and valuable mineral resources being cases in point.

A political state encompassing an entire archipelago is a rarity; the Republic's neighbour Indonesia, for example, would have experienced a less troubled past if the whole of Kalimantan (Borneo), Irian (N. Guinea), and even Timor fell within its boundaries. This unique asset of the Philippines enables the Republic to be both a part of Asia and yet stand aloof from Asian affairs. Just as the British Isles is the least European part of Europe so too the Philippines is the least oriental country in the Orient. Furthermore, the Philippines have been enriched, as have been the British Isles, both by continental influences and by trans-oceanic influences.

The current level of economic, and indeed social, activity in the Philippines can be established by reference to the trend-setter in South and East Asia—Japan. Only two socio-economic indicators are more favourable for the Philippines than for Japan: increase in agricultural output and exports as a percentage of GNP (Table 1). These represent two spheres in which the Philippines, as a developing nation, might well be expected to outshine a country such as Japan whose much more advanced society consumes most of its agricultural output, and indeed most of its output of all goods and services. It is interesting to note, however, that in certain other spheres the performance of the Philippines *vis-à-vis* Japan is meritorious. The number of physicians, for example, at 81 per 100 000 persons and also road density, at 195 kilometres per 1000 square kilometres approaches Japanese standards. The former ratio owes much to the health and educational programmes of the Americans in the first half of the twentieth century; though, ironically, if the more recent migration of Filipino physicians to the United States could have been avoided, an even more favourable situation would have been recorded. In the case of road density, Philippines performance is perhaps exaggerated in that in terms of well surfaced roads Japan has a distinct advantage.

The Philippines performance *vis-à-vis* other nations of South and East Asia has both good and bad points. In terms of investment and savings as related to GNP, literacy level, and

TABLE 1
Comparative Socio Economic Indicators

	India	Indonesia	Japan	Malaysia	Pakistan	PHILIPPINES	South Korea	Sri Lanka	Taiwan	Thailand
1. Per capita GNP (£)	46	42	596	142	46	88	88	79	125	63
2. P.C. GNP (ann. % increase)	1.1	0.8	10.0	3.8	2.9	1.9	6.4	2.1	6.3	4.7
3. Population increase (%)	2.3	2.4	1.0	3.0	2.7	3.1[a]	2.6	2.4	3.0	3.1
4. Life expectancy (years)	50	32	71	65	43	55	61	62	n.a.	50
5. Physicians (per 100 000)	17	4	111	21	14	81	42	22	70	14
6. Radios (per 1000)	17	13	250	52	10	17	70	n.a.	110	70
7. Newspaper circulation (per 1000)	n.a.	8	482	73	5	25	70	43	83	20
8. Literacy rate (%)	28	30	98	43	20	75	71	75	80	70
9. Agric. output increase (index)	147	135	158	208	164	166	199	156	180	219
10. Roads (km per area)	75	11	201	110	21	195	400	525	145	11
11. Electricity (kWh per cap.)	78	19	2110	260	33	134	169	45	550	45
12. Investments (as % of GNP)	16	n.a.	35	19	15	22	18	14	19	23
13. Domestic Savings/GNP (%)	13	n.a.	35	21	10	20	9	12	18	21
14. Cost of living (index)	166	68319	143	105	136	145	245	127	122	117
15. Exports/GNP (%)	5	n.a.	9	42	4	14	6	23	16	16
16. Exports (annual % increase)	3	-1	16	3	9	8	42	-2	24	8

Source: Based on A. D. Redding, 'The Philippine Economy: A Newcomer's Perspective', *Philippine Economic Journal*, Vol. 8, 1969, pp. 132-3.
[a] Official figure but believed to underestimate reality which approached 3.5%

Data are mainly from the following 1968 publications: *Economic Data Book* (AID), *International Financial Statistics* (IMF), and *World Handbook of Political and Social Indicators* (Russett et alia). They have been updated by reference to IBRD and ECAFE statistics and also 1971 Yearbook, *Far Eastern Economic Review*.

1. Per capita GNP (£)-in current 1969 prices. 2. Per capita GNP (annual percentage increase, 1960-69)-at factor cost. 3. Population increase (percentage)-mid 1969. 4. Life expectancy (years)-official estimates for 1965, 1966 or 1967. 5. Physicians (per 100 000 persons)-most recent data (1964-8) except Taiwan (1969). 6. Radios (per 1000 persons)- for years 1966, 1967 or 1968. 7. Newspaper circulation (per 1000 persons)-for years 1966, 1967, 1968 or 1969. 8. Literacy rate (%)-for 1966, or more recent unofficial estimates. 9. Agricultural output change-1970 index of production (1952-6 = 100). 10. Roads (kilometres per 1000 sq. km of territory)-kilometres of improved roads as of 1966 or 1967. 11. Electricity (kWh. per capita)-for 1965, 1966 or 1967. 12. Investment (as a percentage of GNP, in constant prices)-average for 1963-6 or 1963-7 except for Taiwan (1963-5). 13. Domestic savings (as a percentage of GNP, in constant prices)-domestic savings were derived by subtracting the import surplus from, or adding the export surplus to, investment; the surplus, i.e. net foreign balance, was not deflated. Coverage is as for 12. 14. Cost of living-mid 1970 index (1963=100). 15. Commodity exports (as % of GNP, in current prices)-average for five year period with initial year 1963, except for Pakistan (1964), and terminal year 1967, except India, Taiwan and Thailand (1966). 16. Commodity exports (annual % increase)-IMF dollar-value data for 1962 and 1967 were used to compute the annual average (compound) rate of growth.
Note: data for 4, 10 and 15 for Malaysia refer to West Malaysia only and are thus higher than if they referred to all of Malaysia.

electricity consumption, the Republic's record is above average. On the other hand, it is not impressive in respect of population increase. The rate of population increase is perhaps the most critical socio-economic indicator where developing nations are concerned, and that for the Philippines, in excess of 3.0% per annum, leads all other South and East Asian nations.

Contemporary society in the Philippines is thus faced with major problems arising in no small manner from the character of the country's physique and its socio-economic features. This is a situation common amongst the world's developing nations. However, the Philippines also possesses several unique characteristics which complicate the situation: in the context of South-East Asia, it was the first nation to achieve independence from colonial role; it is the only Christian country in all Asia; and it claims to be the third largest English speaking country in the world. With its recent past consisting of over three centuries of colonisation in its more extreme form by the Spanish, and a further half century of enlightened rule by the United States, the Filipino's way of life and livelihood in many respects sets him apart from his neighbours in South-East Asia despite close affinities in terms of racial characteristics and natural resources.

Data Presentation

People more than place form the key to any geographical study of the Republic of the Philippines today. If a reasoned portrayal is to be given, special attention must be paid to the way of life and livelihood of the Filipino. In the past, social and economic elements have been closely entwined and have also been in close rapport with nature. In recent years, and with increasing rapidity, these inter-relationships have become less closely knit, with the role of nature also declining in significance. As the nation continues to develop these trends can be expected to persist. Also, Chisholm's challenge to geographers: '... are we going to be content with the widely held opinion that geography is concerned with the past and the present but leave concern with the future with others?'[1], has been accepted in a small way in this volume, which concludes with an attempt to remedy this deficiency in approach by discussing the prospects for the geography of the Philippines in the foreseeable future.

In presenting and analysing the salient features of the geography of the Philippines, national and regional observations are supplemented, wherever possible, by case studies. Most case studies are to be found in Chapter Eight (The Present) where they serve to highlight or exemplify some of those features considered most pertinent to an appreciation of the nature of the islands' geography. As such, they appear also in Chapter Nine (The Future) for it is contended that those features of today's geographical patterns of most interest must be those which will have a vital role to play in shaping the patterns of tomorrow. Not only do the case studies supplement the detailed reference material which increasingly occupies the later pages of the volume, but, they serve as valuable examples for the student, and they also provide deeper insight into the nature of Philippine geography for the educated layman.[2]

The subject matter of the case studies is diverse. Key features of Philippine geography which are examined include various demographic characteristics; the near insoluble problem of land reform; developments of great future import such as high yielding varieties of rice and the imminent ending of the Laurel-Langley trade agreement; and the scope for industrialisation. Urban geography is especially amenable to case study presentation and a variety of settlements at all levels in the urban hierarchy are so examined. The case study approach is also extended to socio-economic aspects, notably suburbanisation, Filipinisation and, in the field of political geography, to the Sabah issue.

This volume is designed for reference purposes, for the student and the educated layman. The text is so arranged as to lead the reader step by step towards the wide-ranging and penetrating geographical assessments contained in the final chapters. Thus, the volume commences with an introductory verbal picture, to serve as a part-impressionistic, part-factual frame of reference, of the Philippine 'landscape' (Chapters Two and Three—The Urban and Rural Scene respectively) and then proceeds to a comprehensive review of the principal elements of the physical and human environment which have created this landscape (Chapters Four and Five—The Place and The People respectively). Then, using the geographers' special technique, regional analysis, the setting in which the way of life and livelihood is played out is succinctly described and the areal differentiation caused by the gradual variations in the spatial interaction of physical and human elements which are continually taking place is also demonstrated.

After the regional analysis there follows the end product of this geographical study—a scientific description of the archipelago insofar as it is the home of the Filipino nation. Phenomena of relevance are interpreted by means of a geographical appraisal which attempts to demonstrate how the present, impermanent, pattern of life and livelihood has evolved, and hence to predict the modifications most likely to take place in the future. Proper appreciation of the discussion contained in Chapters Seven, Eight and Nine (The Past, Present, and Future respectively) is not possible without the prior orientation and analysis which forms the subject matter of the preceding chapters. The assessment contained in the concluding chapters is streamlined as far as possible through the use, without elaboration, of phenomena discussed in detail in preceding chapters. Past influences modifying geographical patterns are traced with particular reference to those still exerting direct or indirect influence. Current patterns are assessed in detail and those of most significance to the future geography of the Republic are identified. The causal factors responsible for such patterns are established and their probable future impact is assessed; such an examination makes possible informed predictions as to any likely modifications to the existing nature of Philippine geography in the next decade and beyond.

It must always be remembered that, whatever the past and present influence of the Occident, the Philippines is an integral part of the Orient and should become increasingly so. Consequently, current geographical theories and techniques must be employed with caution; concepts arising from Western experience may not necessarily be valid when applied to Asian cultures. The present author does not claim to have succeeded in overcoming these difficulties, but he does possess an important advantage denied to most other Western students of Philippine geography in that family ties have made it possible to live *with* rather than *amongst* the people, both in an urban and in a rural environment. Furthermore, the locales involved were not generally those frequented by visiting academics, which enabled contact with a much broader spectrum of places and communities than usual. This was especially significant in view of the often considerable gap between social levels in the Philippines.

References

1. M. Chisholm, 'The Kingdom of the Blind', *Econ. Geog.*, Vol. 45 (1969), Guest Editorial (January Issue).
2. An admirable factual study of the country is to be found in F.L. Wernstedt and J.E. Spencer, *The Philippine Island World: A Physical, Cultural and Regional Geography*, University of California Press, Berkeley and Los Angeles, 1967. But, like other geographies of the Philippines, it succeeds only in conveying to the reader the details of the phenomena to be found there and generally fails in its attempt to give an impression of the end result of these phenomena—the environment in which the Filipino and Filipina live and work.

2
The Urban Scene

In an underdeveloped country such as the Philippines, one might expect a description of the rural scene to precede that of the urban scene. This procedure has not been adopted in this case in view of the overriding importance of the urban centres, and of the metropolis of Manila in particular. Most events and decisions which shape the pattern of life and livelihood in the Philippines emanate from Manila while the other urban centres, to a greater or lesser extent, provide the outlets by which the consequences of these developments are disseminated to the nation as a whole. The foremost facets of the Filipino culture and development, not to mention a disproportionate segment of the total population, including the bulk of the nation's decision makers, are to be found in the urban centres whose relative liveliness and sophistication contrasts vividly with the routine and poverty of most rural areas.

A veneer of sophistication and oriental charm embraces in varying degrees all the urban centres of the Philippines. Fine buildings, vividly decorated public transport and well groomed businessmen may readily catch the eye[1] but, to the more discerning observer, slum dwellings, potholed streets and low water pressure (if piped water is, in fact, available) are of equal or greater relevance as components of the urban scene. Such marked contrasts typify many aspects of the urban dweller's life and livelihood. Many are universal in application, such as elaborate residences which can only be approached along badly maintained streets; others are the product of regional differences, such as the juxtaposition of traditional Moslem and modern ferroconcrete building styles in the larger towns of Southern and Western Mindanao. Such contrasts are to be found at different levels of society also. Thus, the company executives of the opulent and Americanised suburb of Forbes Park in

Manila form a community very different in character from that of the often equally wealthy, but relatively inhibited and old fashioned, society created by the sugar 'barons' of Bacolod in the Visayas. Similarly, both the Moro dock labourer of Zamboanga City in Mindanao and the Ilocos road worker from Vigan far to the north of Luzon may form equally poor and notoriously violent communities but in organisation the two societies differ radically, since for the former the pivot is the mosque while for the latter it is the church.

To present a comprehensive picture of the urban scene it is necessary to supplement a discussion of the metropolitan scene with discussion of other elements, particularly those identified in Ullman's urban hierarchy.[2] The results cannot fully illustrate the diversity of urban landscapes to be found in the Philippines but it is considered that the six case studies selected—Manila (the national centre), Cebu (an inter-regional centre), Iligan (an industrial centre), Dumaguete (a major provincial centre), Hilongos (a minor provincial centre), and Trinidad (a *barrio*, or village) between them contain most of the principal elements of the urban scene (Fig. 2).

Particular emphasis is placed upon a discussion of Manila and Hilongos. The metropolis of Manila dwarfs all the other urban centres and, therefore, is both atypical and representative of the urban scene in the Philippines. As the traditional entry point to the country, Manila and its immediate environs offer the visitor his first, and often only, impression of the Philippines. In doing so it is possible to gain considerable insight into the nature of the urban scene but in some respects such an impression may be distorted and incomplete. Hilongos, on the other hand, typifies the settlement pattern and way of life best known to the average Filipino, be he country or town dweller.

Local variants to the patterns revealed by the case studies are to be found in parts of the country reflecting regional influences, e.g. the less sophisticated town structures of parts of Mindanao, or local physical features, e.g. ribbon as opposed to nucleated settlement for coastal towns resulting from the nature of the foreshore. However, such variations, interesting as they are to the micro-geographer in revealing the influence of particular environmental elements, distract from the objective of this study which is concerned to identify basic elements of

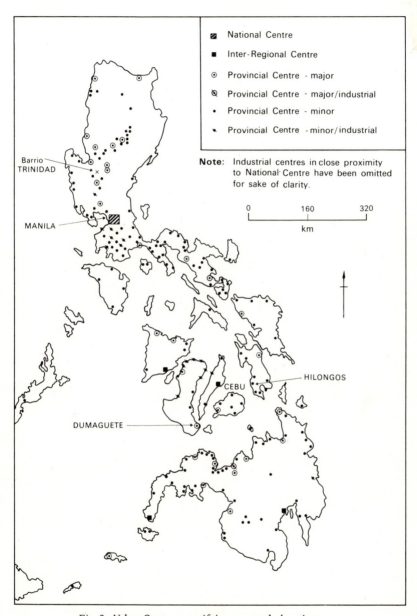

Fig. 2. Urban Centres, specifying case study locations

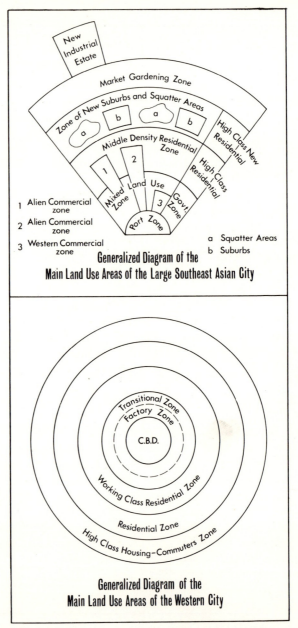

Fig. 3. Generalised diagram of the main economic zones in the Western city and the Southeast Asian city. The upper sketch is applicable, in varying degrees, to both Manila and Cebu

Philippine urban geography which, as McGee[3] has demonstrated, may differ markedly in form and function from those familiar to the overseas visitor (Fig. 3).

1. Manila—the National Centre

Manila is a metropolis: the urban population is currently estimated at 3 million persons or 8% of the total population of the country. Urban development has long since overflowed the boundaries of the City of Manila and currently embraces part or all of more than half a dozen adjacent local government units (Fig. 4). Manila is one of the great cities of the Orient and consequently is organised on a vastly different scale to any other urban centre of the Philippines: for example, it consumes 90% of the nation's electrical energy and houses one half of the nation's manufacturing activity. Such is the preeminence of Manila that the country's second largest city—Cebu—possesses only the equivalent of one quarter of the population of the City of Manila and Cebu itself is 50% larger than Iloilo, which is the third most populous city of the Philippines (Table 2).

The metropolitan character of Manila is immediately evident to the new arrival. The most common mode of approach is by sea or by air; in either case the impressive extent of urban development is revealed. From the sea, Manila appears to occupy the entire eastern horizon of Manila Bay. In itself the skyline is undistinguished for the shore of the bay is lowlying and featureless and associated urban development lacks the grandeur of Hongkong or the oriental atmosphere of Bangkok. The spectacle from the air is similarly impressive yet undistinguished: an aerial approach reveals not only much of the best but also much of the worst of Manila. Lying between the impressive sweep of Manila Bay and the rugged forest-clothed peaks of the Sierra Madre to the east, many spectacular developments are visible, such as the University of the Philippines which has been transferred from its cramped downtown location to a spacious campus in outlying Quezon City (the administratively separate, recently designated, capital of the Philippines), the arrow-like fishing traps in Laguna de Bay, and the high-rise office block complex of Makati. However, these features contrast vividly with the haphazard street alignments,

TABLE 2
Principal Urban Centres

Chartered City Name[a]	Designated Area	1960 (census)	Adjusted[b]	1970 (Estimated)	Adjusted[b]
Angeles	n.a.	75.9	76.0	134.0	134.0
Bacolod	60.6	119.3	93.0	180.4	175.0
Baguio	18.8	50.4	50.0	83.0	83.0
Butuan	209.9	82.4	31.0	116.9	45.0
Cagayan de Oro	159.3	68.3	35.0	94.5	55.0
Caloocan[c]	21.5	145.5	120.0	273.0	250.0
Cavite[c]	4.5	54.9	55.0	77.1	77.0
Cebu	108.4	251.1	225.0	345.0	345.0
Cotabato	67.9	37.5	30.0	51.9	40.0
Dagupan	14.3	63.2	20.0	89.0	70.0
Davao	853.7	225.7	85.0	315.3	150.0
Iligan	282.0	58.4	31.0	82.9	45.0
Iloilo	21.6	151.3	150.0	213.0	213.0
Lapu Lapu	22.4	48.5	40.0	67.8	60.0
Legaspi	59.3	60.6	60.0	74.8	75.0
Lucena	26.4	49.3	28.0	76.7	70.0
Manila	14.3	1138.6	1138.6	1323.4	1323.4
Naga	29.9	55.5	55.0	79.2	79.0
Pasay[c]	5.4	132.7	133.0	205.0	205.0
Quezon[c]	64.1	398.0	380.0	751.0	751.0
San Pablo	82.7	70.7	30.0	99.0	42.5
Tacloban	38.9	53.6	36.0	74.1	50.0
Zamboanga	546.1	131.5	28.0	188.3	40.0

Source: F. L. Wernstedt and J. E. Spencer, *The Philippine Island World: A Physical, Cultural and Regional Geography,* University of California Press, Berkeley and Los Angeles, 1967 and Philippine Bureau of the Census and Statistics: Population Projections (issued Aug. 13, 1969) and Advance Reports of 1970 Census (no date).

[a] Officially, a Chartered City should be designated as such, e.g. Cebu City, Lucena City, etc., but generally this has not been adhered to in the text except where confusion with a province name might result, e.g., Quezon City.

[b] Adjusted to exclude non-urban population.

[c] Essentially part of Metropolitan Manila and thus not a discrete urban centre.

Note: the list excludes certain urban centres because not being Chartered Cities in 1960, available data is incomplete; also some major urban centres still were not Chartered Cities in 1970. In the first category fall Olongapo (est. 1970 population 100 000 persons), Tarlac (50 000), General Santos (40 000), Mandaue (40 000) and Jolo (40 000). In the second category fall Pasig (150 000), San Fernando (80 000) and Malolos (75 000).

the vast areas of overcrowding and, on closer inspection, the traffic-clogged streets and areas of shanty dwellings.

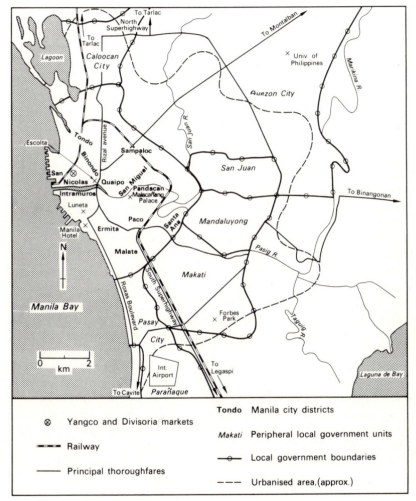

Fig. 4. City of Manila and its environs

The shoreline of Manila Bay, particularly to the south of the Pasig River, serves as a facade behind which are concealed the unsolved problems arising from uncontrolled urban expansion in a tropical climate. The fundamentals of the urban scene are to be found in the suburbs of the metropolis but, although atypical, 'International Manila' as the Manila Bay facade can be described, makes an important contribution to Filipino life and livelihood. Its northern portion, with the narrow thoroughfare of the Escolta as the focal point (Fig. 4) may be regarded as the central business district. The Escolta serves as the principal workplace for the business and commercial interests and the mass media personnel. The latter, such is the near *laissez faire* environment of the Philippines, are able, like their American counterparts, to exert a disproportionate impact. As such they indirectly control and mould the way of life and livelihood of the nation's inhabitants initially in the metropolis but followed soon after in the provinces; widespread literacy aids this process of diffusion which is very rapid by developing nation standards (despite geographical fragmentation, information dissemination and also population mobility in the Philippines are very pronounced).

Across the Pasig River, International Manila extends southwards along another well known highway, Roxas Boulevard. Here, the Embassy of the United States symbolises the American dominance of this part of the metropolis (some would say of the nation as a whole); certainly it provides the focal point of this part of the metropolis both for the American tourist and, due to the presence of many government departments and the head offices of national and international concerns, the American businessman. No longer, however, is the American presence unchallenged; particularly where businessmen are concerned, the Japanese and Europeans vie with wealthy Filipinos in filling the area's luxury hotels and fashionable shops. International Manila is not the exclusive preserve of the wealthy. For the middle class *Manileño* a promenade in the cool of the evening in the Luneta followed by a meal at one of the numerous Bay-side restaurants is not an uncommon experience.

The Luneta is a large park which can be regarded as an offshoot of a district of the metropolis, Intramuros, whose contribution to its life and livelihood lies more in the past than the present but which nevertheless has been of profound significance. Intramuros itself, a quarter of the city only a square kilometre in area, was the original Manila as founded by the Spanish in the sixteenth century. Until razed to the ground in 1944, its quiet checkerwork of streets lined with old houses whose first floors overhung the pavements was dominated by the towers of its many churches. Of these, only two survive today amongst the many as yet undeveloped bomb sites which housed for many years one of the shanty towns of the metropolis.

Prior to the war the principal government institutions were removed from Intramuros beyond the encircling moats of the walled city (which have now been filled in and, in part, serve a much more mundane purpose as a golf course) to a site adjacent to the Luneta. At the turn of the century, these open spaces served a far less peaceful purpose for it was here that the national hero of the Philippines, José Rizal, met his death before a firing squad. Whilst to the Filipinos Rizal's memory is forever green, the Luneta itself, like so much of Manila, remained neglected and unkempt for over half a century. Today, however, it has undergone a transformation into the New Luneta, a formal park not only attractive but also clean, a veritable miracle for the metropolis.

That the New Luneta is an exceptional feature of the urban landscape is self evident from even a cursory inspection of suburbia in the metropolis. The *peso* millionaires enclave of Forbes Park apart, in few locales is there to be found any concerted movement toward pleasant, let alone gracious living: the bougainvillaea cutting sprouting luxuriantly out of a rusty tin can symbolises the haphazard nature of suburban living.

Before considering suburban Manila, attention must be turned to International Manila's neighbour to the north. Whereas to the south of International Manila urban development merges into suburbia, to the north lies the home and/or work place of three sections of Manila society who together pose more problems to the city's administration than the rest of its inhabitants put together—the Chinese, the students, and the slum dwellers of Tondo. Each of these three elements of Manila society has cause, to a greater or lesser degree, to regard themselves as under-privileged. However, the pros and cons of this situation have their roots in politics and associated fields; only indirectly is geography involved.

Student activity is focussed in this part of the metropolis for it contains two of the leading Philippine schools of learning, the

Santo Tomas and Far Eastern Universities, around which have proliferated ancillary institutions. The focal point for their growth has been the University of Santo Tomas founded by the Dominicans in 1611; the incentive for growth has been the ever increasing number of knowledge-hungry Filipinos and, in the case of the all too frequent 'diploma mill', the freedom from non-profit-making restraints where private tertiary education is concerned. Chronic overcrowding of the area now is commonplace but so far only the state-aided University of the Philippines has been able to escape to the relative seclusion of Quezon City.

Student overcrowding pales into insignificance to the northwest, in Tondo. Here lies the metropolis's greatest concentration of the least fortunate, the least able, and least willing to work amongst the population of Manila. Originally, this unsalubrious semi-swamp served as the dwelling place of port workers, market porters and other unskilled labourers. Population growth in the metropolis has established in the area, through its low rent yet accessibility to port-oriented industries, to the central business district, and to International Manila in general, a much broader spectrum of inhabitants which now encompasses most categories of lower-paid workers.

A further hazard of the Tondo environment is fire; early in 1971, for example, some 70 000 persons were rendered homeless in a matter of hours. Major fires are a regular ocurrence in Manila and in other Philippine urban centres, both large and small. Tondo is particularly susceptible to fire because of its bay side situation exposed to the strong on-shore breeze generated by the temperature island formed by Manila in the heat wave conditions which precede the monsoon rains. Any fire fanned by this breeze is difficult to control, for Tondo, like other heavily populated areas of poverty in Philippine cities, consists of homes huddled closely together and made of easily combustible materials, while in some sections streets are virtually non-existent, making access difficult.

Although data is not available it is quite possible that over-crowding in Tondo is surpassed by that of the area which lies between it and the central business district. This is Chinatown, a compact quarter of narrow crooked streets containing some 100 000 Chinese. Its refuse-laden canals have gained Chinatown the nickname of Little Venice. A recent upsurge in anti-Chinese feeling has increased the ghetto-like atmosphere of the area. The vitality and colour of Little Venice has been muted by the forced removal of the often flamboyant Chinese language signs, the curtailment of retail trading, and the continued displacement by motorised transport of the traditional *calesa*. But, despite increasing pressure, the Chinese continue to exert considerable influence in business circles: wholesale trade is Chinese dominated and in the innumerable backstreet factories of Chinatown are still produced the ready-made trousers, the colourful shirts, and Woolworth-type articles which find a ready market, particularly amongst the less affluent inhabitants of the metropolis.

For the inhabitant of Tondo, shopping in Manila's Oxford Street (Rizal Avenue, which runs north from the eastern end of Escolta) is for special occasions only; his purchases normally are confined to the neighbourhood market. One of the best known and most important of Manila's markets is located on the Tondo-Chinatown boundary. This is the Divisoria market which, together with the nearby Yangco textile market and associated retail stores, constitutes a shopping complex of truly metropolitan dimensions.[4] Such is the hinterland of the market that over 10 000 vehicles are believed to serve it daily and in consequence there has evolved in its vicinity Manila's largest public transport terminal. Within the Divisoria market several thousand stalls display an often bewildering array of fresh foods, groceries, beverages, household items, toys and confectionery. Around the market, and as close as possible to it, crowd a motley collection of parasitic retail stores and stalls offering practically any object imaginable. Clothing and other textiles are especially popular as they are forbidden in public markets such as Divisoria, though not at private markets such as Yangco (Chinese dominance of the texile trade and the reservation of stalls in public markets for Filipino traders combine to bring about this state of affairs). One consequence is that most parasitic retail outlets are Chinese owned, though no longer overtly operated as such as a result of legislation designed to reduce the dominance of the Chinese in business.

Regular patrons of the shops in Rizal Avenue are the middle class inhabitants of Manila. But, despite the growth of employment opportunities in industry in Manila, they still are small in numbers and in influence. Any individual, given the energy, the initial capital and the ambition can live (usually ensconsed behind the tall broken-glass-topped garden walls necessary to protect both person and possessions) in a considerable measure of comfort with his servants, his TV set, automobile and air-conditioning, and entertain generously as well. It is to secure these rewards that thousands of immigrants descend annually upon the metropolis. For most they prove illusory objectives and the newcomers merely swell the already unmanageable population which illegally squats on open spaces as close as possible to places of employment, notably the Port and the factory concentrations dotted throughout the metropolis, or else on the fringe of suburban subdivisions.

While the scope for free enterprise is unchecked, community development is thwarted on all sides by the inertia and incompetence which characterises most facets of local government and associated community activity. This state of affairs is brought about by a combination of vested interests, corruption, indifference, the magnitude of the problem and the absence of initiative at a national level. Western patterns of urban growth are thus distorted and the main characteristics of the suburban scene include shanty towns and squatters, congestion and chaos (one finds houses built in the middle of gazetted roads), corruption and 'carnapping', and garbage piles and gang warfare. In this context it should not be forgotten that the metropolis represents one extreme of the urban scene in the Philippines; for example, the crime rate in Manila is almost fifty times greater than that for the rest of the country.

Manila is a metropolis of many facets. Those touched upon above will, it is hoped, have provided the reader with an indication of the principal features and problems of the national centre of the Philippines. These are discussed in greater detail when the time comes for regional analysis, for its uniqueness and importance warrant the treatment of Manila as a regional entity, notwithstanding its limited areal extent. Furthermore, in subsequent chapters certain aspects of the urban geography of Manila will be examined in greater detail. Receiving particular attention are those aspects, notably suburbanisation, likely to have a profound impact upon other Philippine urban centres in the future, and indeed upon the economic and social geography of the country as a whole.

2. Hilongos—a Minor Provincial Centre

In complete contrast to Manila is Hilongos, one of over a hundred minor provincial centres which can be distinguished in the Philippines. The minor provincial centre serves a restricted, usually agricultural, hinterland for many of whose inhabitants it is the only urban centre they will ever visit. This hinterland generally is well defined by topographical features and, in consequence, intercommunication between minor provincial centres and between the major provincial centre is often handicapped, thus increasing the insularity of the former. Both in form and function the minor provincial centre is essentially a miniature version of the major centre except that its limited hinterland precludes the establishment of colleges, insurance offices, bus depots and similar manifestations of a regional, as opposed to a local, sphere of influence. Normally, it is the smallest settlement unit in which can be found dentists, lawyers and other such specialised tertiary activities.

Hilongos, with almost 5000 inhabitants in 1970, ranks amongst the largest of the urban centres of the island of Leyte and in 1959 the town's income enabled the municipality to gain first class status. The town acts as a service centre for the largest area of alluvial soil in South-Western Leyte, the Salog River floodplain. It is located, as are the majority of such centres, where the means of intercommunication, in this case the coast road, meets the access route to the hinterland, in this case a road parallel to the Salog River. The settlement pattern of Hilongos is in many ways a microcosm of the indigenous features of urbanisation in the Philippines. Outstanding is the space taken up by individual urban lots. This traditional Malay pattern owes its present-day persistence in no small manner to its effectiveness in controlling the all too frequent and often disastrous outbreaks of fire. Only in the central portions of urban settlements and in Manila where subdivision is sufficiently advanced does this rural aspect of the urban scene disappear. The result is a very mixed form of development: in the better residential areas, for example, outbuildings containing servants' quarters may occur, while the greater part of the urban area is in fact given to coconut groves and rice fields upon which squatters' homes add further to the complexity of urban development.

Even in an insignificant urban centre such as Hilongos, it is possible to distinguish relatively homogenous residential areas, categorised as superior, inferior and central (Fig. 5).

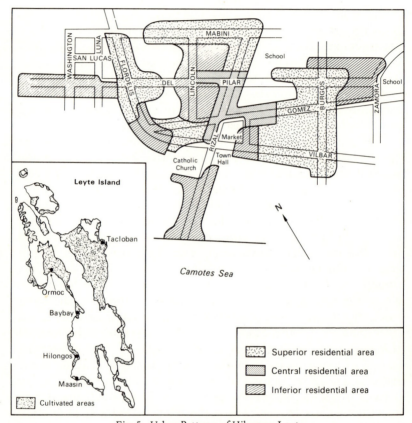

Fig. 5. Urban Patterns of Hilongos, Leyte

Superior Residential Areas. The principal concentration of superior residential accommodation in Hilongos is located at the eastern end of Rizal Street. In this block almost all dwellings are of two storeys and exceed 200 square metres in floor space. Wooden construction predominates, sometimes supplemented by a concrete/breeze block lower storey. Other dwellings in the area are more commonly a little smaller in size (150-200 square metres) than these 'mansions'. Two sections can claim to be the next most exclusive housing areas in Hilongos. The larger has Burgos Street as its focal point; the other smaller section lies athwart the eastern section of Flordelis Street. Blocks of two storey houses again predominate but concrete is less common as a building material and floor areas drop to the near 100 square metres mark. The 1970 Census preliminary data suggests that about one fifth of all dwellings were 'strongly constructed': contrast this with 'more civilised' Dumaguete (two fifths), Cebu (47.5%) and Manila (95%).

The three superior residential areas of Hilongos form a crescent enclosing the town centre on all but its western side. This development would appear to be influenced by site conditions, and access to the town centre and place of employment respectively. In terms of site, the crescent generally is aligned along slightly higher terrain which consequently is relatively less marshy, flood prone and mosquito ridden (the two breaks in the crescent in fact indicate the presence of stream channels with associated marshy areas). The crescent is strategically placed with respect to the town centre which dominates the social and economic life of the settlement and its environs, while a major traffic route (that which parallels the Salog River) passes through the Flordelis section. In contrast to this ease of access to the principal agricultural areas, the Burgos section is oriented to the sea and it is also in close proximity to the high school. The eastern Rizal section has no transportational advantage, which in terms of lack of disturbance is beneficial. It is, however, located adjacent to the elementary school.

As Hilongos is primarily an agricultural service centre its superior residential areas contain the homes of numerous rice, maize and copra farmers and merchants (most of whom are also substantial landowners, i.e., possessing five or more hectares most of which will be operated on a tenancy basis). Their presence is not difficult to spot: a rice or maize mill may form part of the dwelling complex and storage space for these and other agricultural products is commonplace. In many cases the entire ground floor is devoted to business activity with the garden often given over to storage, while the edge of the road is utilised for drying copra and other agricultural products.

Individual superior areas also attract their own distinctive inhabitants. The presence of the sea and the high school attracts both school teachers and fish merchants to the Burgos area while its proximity to the town hall may also account for the presence of several notaries public. The Flordelis section on the other hand has a mainly agricultural flavour. In eastern Rizal Street school teachers again figure prominently amongst householders, the adjoining elementary school being a contributory factor.

Central Residential Areas. Residential accommodation is an essential element of the town centre whose main axis is Gomez Street but which possesses a subsidiary grouping of business premises, linked by central Rizal Street, at its intersection with the coast road (del Pilar street). Dwellings are usually of relatively high quality; in most cases residences are positioned over business premises (shophouses) and are constructed of either wood or concrete. Floor spaces are around the 100-150 square metres mark. Residents, who include the majority of the town's handful of Chinese, usually comprise the family which owns or rents the premises below. Such tradespeople are among the wealthiest inhabitants of Hilongos as agricultural interests (both through trade and land ownership) provide valuable additional sources of income. Convenience, plus the desire to be on the spot to protect the premises' contents from the ever present dangers of fire and theft, are the dominant locational factors for such accommodation.

Residential accommodation, of course, is only a small part of the complex of structures that make up the town centre. The most imposing feature is probably the church but the most significant is undoubtedly the market itself, a series of shed-like structures under whose roofs, and in the intervening open spaces, the basic function of Hilongos—trade—is performed, together with recreational activities such as basketball, billiards and, formerly, cock fighting. The junction of Rizal and Gomez Streets can be regarded as the centre of the town: here, for example, are located the principal retail stores and the advertisements for film shows; it is the bus stop, the terminus for the *jeepney* (including a motorcycle powered version which is superseding the traditional *tartanilla*) and other modes of public transport and the meeting place of local gossips. Overshadowed by the church and the market are the town hall and Association of *Barrio* Councils Centre, buildings which should become of increasing importance once government attempts to improve social and economic conditions bear fruit. At present, however, the public square in front of the town hall vies for prominence as politicians outbid each other with promises of what they would do if they were allowed to run the town hall.

Inferior Residential Areas. Western Rizal street is in complete contrast to its eastern portion: homes with less than fifty square metres of living space predominate; only a few exceed 100 square metres. Closely packed single storey structures are the rule. As Cutshall succinctly puts it, the building frame is based upon:

'bamboo posts with woven split-ratten (*sawali*) sidewalls, a *nipa* thatch or galvanised iron roof, and split-bamboo floor. Its crowded rooms and rectangular openings in lieu of windows permit the details of family living to become neighbourhood information. The shouts, screams, yelps and squeals of barrio children and livestock are a part of the scene. The smells of the charcoal stove, garlic seasoning and possibly a nearby open sewer lend further atmosphere.'[5]

A similar *barrio* style environment is to be found clinging to the banks of a distributary of the Salog River to the north-east of the church and on marshy ground to the south of the town hall. Poor quality homes stretch westward along Rizal Street right to the wharf (the Pantalan section). Here, as in other areas of similar housing, dwellings are usually built on insubstantial looking stilts to offset the extreme swampiness and, in places, tidal nature of the terrain. Fortunately, here as elsewhere in Hilongos, abundant supplies of artesian water are available so the danger of pollution is reduced though lack of adequate sewerage largely offsets this advantage. The Pantalan section, like the eastern perimeter of the town, enjoys the benefit of firmer foundations, provided in this case by the beach ridge. Offsetting the benefits of this drier site is the Pantalan's increased vulnerability to typhoons which have been responsible for the gap in housing development immediately to the east of the beach ridge.

Inferior residential accommodation in a different environment but of essentially similar morphology to that associated with western Rizal Street is to be found clinging to the outer perimeter of the crescent of superior residential areas and also on a smaller scale within it. In the case of the latter section, which is aligned along Lincoln Street, marshy conditions are the principal locational factor. In the case of the crescent's outer perimeter two areas predominate—at opposite extremities of del Pilar Street. The locational factors responsible for this poor quality housing and its associated 'rurban' extensions are twofold: the desire of these predominantly agricultural workers to live close to their place of employment, or their inability (based largely on economic grounds) to secure a more advantageous homesite closer to the centre of Hilongos. In certain respects, notably a less unhealthy site and the opportunity for a vegetable garden, the inhabitants are more fortunate than their counterparts in the vicinity of western Rizal Street. Offsetting this they are more remote from the centre of Hilongos and have less opportunity to secure cheap fish supplies. Furthermore, the fishermen who dominate the occupational structure of the western Rizal Street section have had an additional source of income from ship to shore porterage since the breaching of the wharf by a typhoon in 1949 (the wharf recently became operational again but the resultant growth of traffic has maintained the level of opportunities for casual employment).

The residential patterns of Hilongos have been discussed at some length for, with such a relatively simple urban morphology, Hilongos more readily serves as an example of urban zonation whose principal features are repeated in more complex form in the country's major provincial centres, in the inter-regional centres, and in parts of the metropolis. It needs to be stressed that the locational factors indicated as influencing the location of superior residences (and by definition thus serving as a frame of reference for all urban dwellers in residence selection) should not necessarily be regarded as being decisive. Neither singly nor in conjunction do they necessarily provide a potent control settlement which, in such a small community, cannot be expected to crystallise into a well defined pattern.

Hilongos is not only a place of residence for its inhabitants but generally it is a place of work also. Some indication of occupational patterns is contained in the survey of residential areas and the predominance of tertiary employment is self-evident. The *raison d'être* for Hilongos, as for most other minor provincial centres, is trade. Such centres form a vital link in the distribution chain which enables the primary producer to dispose of his output and, with the income so gained, purchase such consumer goods and services as his means and desires dictate. Commodity marketing and retail trade are thus the cornerstones of economic activity in rural service centres like Hilongos. In a developing country such as the Philippines these activities are of profound national significance also. Therefore, discussion of their form and function, with special reference to Hilongos, is singled out for case study treatment in the concluding chapters.

Despite having a vital role to play in the commerce of the nation, insularity is the keynote of the urban scene as far as many minor provincial centres are concerned. Constraints placed by terrain are contributary factors but, with an economy almost totally dependent upon agriculture, events in the hinterland are of paramount importance to inhabitants of rural service centres. Paralleling this inward-looking economic attitude, social intercourse revolves very much around local gossip, particularly that concerning the network of relatives established within—rather than without—the hinterland as a result of restricted social mobility.

Hilongos is less inward-looking than some in view of its proximity to the inter-regional centre of Cebu (Hilongos is the port of Leyte closest to Cebu and also possesses the only regularly used airstrip in Southern Leyte). Nonetheless, the majority of inhabitants have limited personal contact with Cebu and even less practical experience of those features of urban life which are taken for granted there. In a town which lacks a telephone service and daily newspaper delivery, the transistor radio and the cinema show are the principal means by which most inhabitants are exposed regularly to the world at large.

Only twice a year is the routine of life in Hilongos disturbed. This is at Christmas and *fiesta* time when sedateness gives way to uninhibited enthusiasm and insularity is breached by an influx of relatives and friends. In many cases the relatives have come from the countryside or adjacent minor provincial centres but, as the drift to the towns accelerates, a significant proportion make their way from the larger urban centres and in particular Manila. With rising incomes and improved communications, the exposure of urban centres such as Hilongos to such relatively sophisticated members of society will become an annual event no longer. For the present, however, and particularly in terms of social intercourse, the minor provincial centre will in large measure remain aloof from the world around it.

3. Cebu—an Inter-regional Centre

The city of Cebu is situated on a narrow alluvial plain at the foot of the central limestone range which traverses the entire length of Cebu Island. Offshore lies coral-fringed Mactan Island where the explorer, Ferdinand Magellan, met his death in 1521. A subsequent expedition under Miguel Lopez de Legaspi established Cebu as the base for the Spanish conquest of the Philippines. Initially, Cebu was the seat of Spanish colonial government and, although this was transferred to Manila in 1571, the city remained a stronghold of Spanish economic interests and prestige. Cebu rapidly came to exert considerable economic and political influence over much of the central and southern Philippines, and in particular, over the Central Visayas.

Cebu is the prime example of an inter-regional centre in the Philippines. Other centres include such well known cities as Davao and Iloilo. With a population of 345 000 persons it is the largest of these centres and in consequence most clearly demonstrates their characteristic features. Despite its size, Cebu has yet to acquire many of the trappings or atmosphere of a major urban centre. The city does possess many of the features of metropolitan Manila but it lacks, for example, the glamour and glitter of International Manila and the impressive architectural complexes of the central government. On the other hand, Cebu retains the sedateness, the intimacy and the inter-relationship with the countryside which are so characteristic of the provincial centres.

Its size and nodal position make Cebu an important administrative, industrial and educational centre but its function is primarily commercial. Cebu in particular resembles a small edition of the port of Singapore, being engaged in the handling, forwarding, reshipping, breaking bulk, sorting, grading, processing, distributing and collecting of both raw materials of its hinterlands and the manufactured goods required by their inhabitants. The port is thus the focal point of the urban scene, attracting the major public markets and the central business

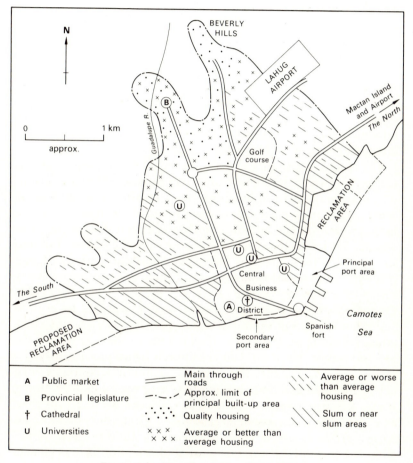

Fig. 6. Cebu City: Sketch Map of Urban Pattern

district (Fig. 6). The latter has been almost completely rebuilt having suffered considerable damage from both Americans and Japanese in World War II.

In the commercial core of Cebu many features of the metropolitan scene are repeated although on a smaller or muted scale. Chinatown, for example, occupies but a few blocks and the impressive fortifications of Intramuros are replaced by a modest Spanish fort. Traffic jams, slums, crime and corruption are all to be found in Cebu but these and other examples of excessive and uncontrolled urbanisation rarely attain the grandiose proportions so characteristic of the metropolis. Suburbia is often chaotic in appearance and, like the metropolis, lacks a solid core of middle-class citizens. Nevertheless, a more pronounced spirit of neighbourliness prevails, supplemented by the more pleasing appearance of all but the innermost suburbs, which is created by a prolific growth of vegetation in turn reflecting the less cramped living conditions and the absence of a long dry season.

The urban scene in Cebu is one in which a compromise is reached between the sophistication and often artificial gaiety of the metropolis and the uncomplicated routine of the provincial centre. One may patronise restaurants, clubs and stores whose facilities are on a par with the better-class establishments in Manila. Their clientèle will not of course be the élite of Filipino society but they are far from provincial in outlook, being graduates of San Carlos or another of the city's four universities, and decision makers and trendsetters in their own right. than, say, with matters of national import but will be erudite and stimulating. On the other hand, for the majority of Cebu's inhabitants life still follows the familiar routine in which the *sari sari* store, the *plaza*, and the church figure prominently. To many, unused to the trappings of modern technology, the newly installed escalator leading to one of the main department stores was first regarded more as an object of wonderment than as a convenient means of access to an establishment most of whose products are beyond the reach of their pocket. Generally, in Cebu only a smattering of sophistication is in evidence. Thus, the prosperous landowner or merchant, and also the successful businessman may live in a luxurious ranch-style house but it will be only one of a handful so far built on the spacious but largely barren piece of speculative real estate euphemistically named Beverly Hills. Furthermore, to reach the jet which will whisk him to Manila in barely an hour his chauffeur-driven limousine will have to board a rusty Second World War tank landing craft which still serves as a ferry, pending completion of the two kilometre long Mactan Bridge.

In part these anomalies are symptomatic of urban growth which throughout the Philippines is increasingly expanding the metropolitan content of the urban scene. This is particularly the case with Cebu, which is especially susceptible to innovations on account of its trading links bringing in not only goods but also people and ideas. At present, for example, the city is engaged in an ambitious port expansion scheme initially involving the reclamation of 160 hectares, whose associated industrial and commercial developments could well transform the urban scene. Then, the horse drawn *tartanilla* will no longer compete with the taxi and the small town atmosphere will linger only in parts of suburbia. For the moment, however, Cebu typifies conditions for the million or so inhabitants of the nation's inter-regional centres, cities in which urban life is a pleasure rather than a necessary evil.

4. Dumaguete—a Major Provincial Centre

Founded by the Spanish in 1534, Dumaguete is the most southerly urban centre of any size in the Visayas. It is located upon one of the larger of the small, isolated and discontinuous segments of coastal plain which fringe Eastern and Southern Negros. The coastal strip, at most only a mile or so in width, is succeeded inland by a significantly hilly area which represents the fringe of the rugged Southern Highlands whose volcanic peaks, approaching two thousand metres in elevation, dominate the landscape.

Dumaguete is the largest urban settlement in Negros Oriental (1970 estimated urban population 25 000 persons) and can be regarded as being representative of the seventy-odd major provincial centres scattered throughout the Philippines. Most, like Dumaguete, have a coastal location and serve a predominantly rural hinterland (also, in the case of Dumaguete, the adjacent small island of Siquijor). In addition to their maritime and commercial functions, the major provincial centres usually act as centres of communications, education, finance, government, recreation and religion. Dumaguete, for example, is the focus of the local road system, possesses a wharf serving more than a dozen vessels per week, and is connected to the principal cities and towns of the Philippines by air, telephone, telegraph and radio. Dumaguete is exceptional in possessing an important university (Silliman) but its possession of other establishments of tertiary education (four colleges) is typical of a major provincial centre. (In total, the student population is about 10 000 of which only 1500 are permanent residents of the city and hence were included in the 1960 Census total.) So too are the various branches of national banks and insurance companies, its provincial legislature, its group of cinemas, and its cathedral (Fig. 7).

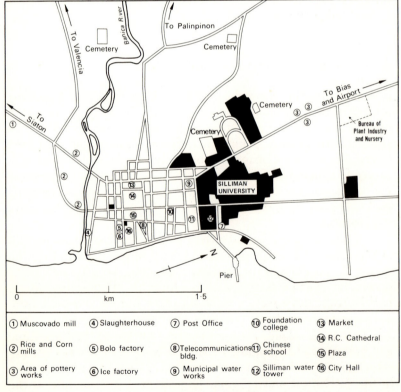

Fig. 7. Dumaguete City

Although Dumaguete was designated a city in 1948, and thus established administrative autonomy from the province of Negros Oriental, it is to all intents and purposes a medium sized town. Many streets are now paved but others have only a graded earth surface, and an often badly maintained surface at that. These roads form a grid pattern for which the universal triumvirate of the *plaza*, the market and the cathedral form the nodal point. Most of the built-up area is within a kilometre of this nodal point so the intimate atmosphere of the small town is well preserved. The commercial centre, too, reflects the absence of any high degree of urbanisation. Both the retailing and warehousing areas are ill defined; typical is a pattern of interspersed business structures and residences, most often a common building with the living quarters above or behind the shop or warehouse. Manufacturing is largely absent, being confined to cottage industries such as pottery making and basic activities such as rice and corn milling and the slaughterhouse.

Despite the small scale of business operations, the world of commerce is still dominated by the Chinese. In numbers their total is quite small (only two per cent of the population in 1960), but virtually all enterprises of any size are financed and also usually operated by the Chinese. If a Filipino wishes to engage in commerce on his own account he finds himself, despite recent discriminatory legislation, able to do little more than run a *sari sari* store, a tailoring business or a barbershop.

However, the Filipino comes into his own in the non-commercial world. For a country where influence in political circles is essential for the success of free enterprise, the Filipino staff of the provincial and city authorities in particular wield considerable power. This is all the more so as the highly centralised administrative structure gives the bureaucrats and politicians in Dumaguete control over all the numerous government agencies operating in the other urban centres of the province (the more northerly of which, due to transport difficulties, fall beyond the boundaries of the city's trade hinterland).

Officials in the upper echelons of government together with members of associated professions, such as the law, enhance the quality of the urban population. Many of the larger landowners in the province also maintain town residences so that a significant nucleus of decision makers, albeit of restricted influence, form the core of local society. In Dumaguete this nucleus is augmented by the staff of the University, the hospital, and the various colleges. The University's mixed American and Filipino staff can be regarded as trendsetters of considerable local significance, notably in the field of home construction, sanitation and cultural activities (it is of interest to note that Dumaguete possessed at the time of the 1970 Census an exceptionally high literacy rate—95%—doubtless influenced by the educational functions of the city). The suburban pattern reflects the influence of Silliman University in that the superior residential area which extends from around the campus along the seashore is larger than in major provincial centres of comparable size. In general, the suburban scene is typical, however, with an atmosphere as much 'rurban' as suburban resulting from the Filipino preference for a detached dwelling, where possible set in a patch of land large enough to raise a few pigs or chickens and to cultivate some vegetables or fruit. Increasing population pressure (the population of Dumaguete city doubled in the period 1948-1960 compared with an increase of less than one third for adjacent rural areas), plot subdivision to accommodate relatives and married children, and illegal but nonetheless successful squatting by the homeless, all combine to gradually eliminate the rural element in the older suburbs.

As yet this phenomenon is more characteristic of the larger urban centres of the Philippines but is also felt in the major provincial centres. Dumaguete and other major provincial centres have yet to experience the full effects of excessive urbanisation. Life remains largely routine and uncomplicated; one is aware of the outside world but largely unaffected by it. How long the intimacy and sedateness of urban life will be preserved is a matter for conjecture. Some centres, notably those with easy access to Manila and other cities, may soon succumb, others in more distant and less developed parts, such as Samar, could well be little altered in the foreseeable future.

5. Trinidad—a Barrio[6]

Lower in the urban hierarchy even than a minor provincial centre such as Hilongos is the *barrio*, or village. Trinidad, therefore, represents the basic element of the Philippine settlement pattern, apart from the *sitio*, or housing group. The socio-economic structure of Trinidad itself is strongly influenced by local land tenure arrangements and as such typifies only a sector, albeit an important one, of the Philippines landscape. On the other hand, the basic features of the life and livelihood of the inhabitants of Trinidad can be regarded as portraying those to be found in many such settlements scattered throughout the archipelago.

Trinidad is located some 150 kilometres north of Manila in a position close to the centre of the Central Plain of Luzon (Fig. 8).

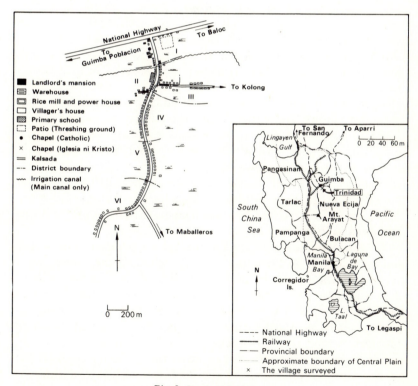

Fig. 8. *Barrio* Trinidad

It forms part of the municipality of Guimba in Neuva Ecija province. The *hacienda* within which Trinidad is located occupies approximately 1000 hectares just to the east of Guimba *poblacion*. It contains within its boundaries three villages—Trinidad, Kolong and Maballero—of which Trinidad is the largest with a population in 1967 of just over 1000 persons, being members of 164 households. As Trinidad is an integral part of a *hacienda*, the villagers are landless: every householder is either a tenant or a farm labourer.

The *barrio* presents a pronounced pattern of ribbon development aligned along the *hacienda's* principal road. At the road's junction with the National Highway is located the mansion of the landlord, or *haciendero*. He, however, rarely occupies it for he has agricultural holdings elsewhere in Luzon together with an important business in Manila where he normally resides. Aside from the *haciendero's* mansion, houses in the *barrio* generally are constructed of bamboo and *nipa* palm, with the better-off inhabitants prefering a wooden structure with a galvanised iron roof. Facilities such as lighting (kerosene lamps) and water supply (hand pumps) leave much to be desired, although improvements to drainage facilities and house lot beautification have taken place recently. Nine *sari sari* stores serve the villagers; their range of goods is limited (the nearest market and specialist shops are three kilometres away in Guimba *poblacion*), but the stores serve also as a focal point of social intercourse in the cool of the evening. In 1967 no household had access to a daily newspaper though many possessed transistor radios; for other forms of entertainment (the cinema or cockfighting) a visit to the *poblacion* is required. Religion, however, is catered for by two chapels—one Catholic and one *Iglesia ni Kristo*.

Educational facilities are confined to a primary school, the largest structure in the village (notwithstanding the *haciendero's* large warehouse and defunct rice mill, and also the *barrio* hall). The rudimentary urban morphology of the *barrio* does not permit the existence of a *plaza* but the threshing ground, or *patio*, serves as a substitute. Life and livelihood in Trinidad revolves around rice. In other parts of the Philippines it equally could be sugar, tobacco, fish or any other primary product for at the *barrio* level man and the land are in close communion. This is no more so than on a *hacienda* where the socio-economic structure leaves the *barrio* inhabitants little scope for anything else. Being both landless and occupying valuable alluvial soil, the villagers' house lots generally are too small to serve as fruit and vegetable gardens as is the case in more favoured locales. Trinidad's inhabitants have made some effort to grow vegetables on waste land in the south of the *hacienda*, and fishing in a nearby stream enables them to seek both sustenance and relaxation.

Social structure in Trinidad is conditioned by the relationship of the villagers to the *hacienda*. This takes three forms— registered tenants, non-registered tenants and squatters. With the *haciendero* an absentee landlord (even his manager, or *encargado*, resides in the adjacent *poblacion*), those inhabitants of the *barrio* with the greatest social standing are the *katiwalas* who exercise supervisory power on behalf of the *encargado* over all the tenants. To the *katiwalas* (five in number) must be added the dozen or so non-cultivating tenants, i.e. those persons who, generally by virtue of other income sources such as a *sari sari* store, are able to sub-let the land to which they are formally registered in their name as tenants with the *haciendero*. The middle stratum of *barrio* society consists of farm operators, comprising both registered and non-registered tenants. Approximately half the *barrio* households fall into this category and, as such, form the core of village society and that which is the most active in village affairs. The balance of householders, nearly 40% of the total, make up the lowest level of *barrio* society—the squatters, i.e. those having no formal entitlement whatsoever to the land on which they reside. Most squatters are farm labourers but included are former tenants and married sons of tenants who not only retain certain possibilities of moving upwards in the social scale but also, and in contrast to the squatters, can be regarded as integral members of *barrio* society through kinship and associated links.

Although the *barrio* represents the smallest settlement unit in the Philippines, social stratification is still present, even though few other features of urban geography are well defined. Trinidad serves as a good example in this respect, for not only does it reveal in embryonic form the basic elements of Philippine society but, being a *hacienda barrio*, such elements are quite sharply differentiated. Five categories of society are present: the upper class, represented by the *haciendero*; the middle class, represented by the *encargado*, and three sub-classes of the lower class, which reflect in a microcosm the threefold division of Philippine society as a whole. These sub-classes are respectively the members of *barrio* society holding a controlling or exploiting relationship with other members, the farm operators, and the farm labourers. Once again it must be stressed that the above patterns are subject to modification in other parts of the Philippines; nevertheless, this glimpse of the urban scene at the micro-level, particularly when compared and contrasted with the situation for Hilongos, provides valuable insight into the more rural aspects of settlement patterns.

6. Iligan—an Industrial Centre[7]

The final type of urban settlement whose characteristics are reviewed in order to provide a comprehensive picture of the urban scene of the Philippines is the most recent to emerge. Its interest lies in the fact that its present prime function has not only superseded the settlement's former role—as a major provincial centre—but also that it is the forerunner of an important new element in the urban hierarchy. As will be discussed in more detail in a later chapter, industrialisation only now is beginning to make a major impact upon the urban geography of the Philippines. In the case of Iligan, however, and also in other smaller settlements where substantial raw material processing facilities are to be found, notably the sugar mill towns of Negros, industry has come to dominate the urban scene.

Iligan received its charter as a city in 1960, some 400 years after its creation as the earliest Christian settlement in Mindanao. The city, the capital of Lanao del Norte, occupies a flat coastal location but immediately inland the terrain becomes increasingly irregular leading to the mountains which mark the edge of the Bukidnon-Lanao plateau (Fig. 9). The urban core occupied 640 hectares in 1967, of which almost a quarter was utilised for industrial purposes (rising to one third if vacant land is excluded). The area devoted to industry is second only to residential land (by a mere 6·50 hectares) as the largest single user of land. Furthermore, most of the newer and larger industrial enterprises have been set up on land outside the urban core, although many are still within the city limits (these embrace an area more than five times the size of the urban core) and certainly within the municipal boundaries.

Industry has been attracted to Iligan through the presence nearby of the Maria Christina Falls on the River Agus whose hydroelectric potential (the river falls 700 metres to the sea in

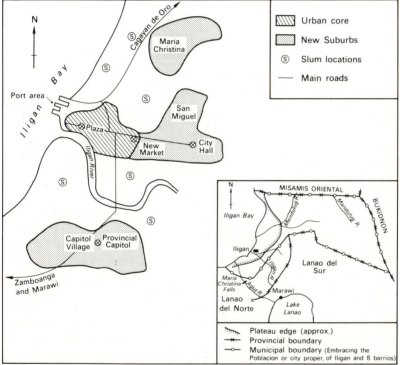

Fig. 9. Iligan City (*Poblacion*) and Adjacent Urban Development.

thirty-five kilometres) is being exploited at seven sites. When fully tapped the power capability of the Falls should total 750 000 kW. Most developments have occurred in the past decade and already have eclipsed Iligan's former functions as the principal port for both Lanao del Norte and Lanao del Sur and as the administrative centre of the former province. In 1948 the city's population stood at 27 725 persons, by 1960 it had reached 58 433 (31 000 persons in the urban core) and then by 1967 it approached 75 000 persons. By 1975 it is expected to reach 100 000 persons.

The boom-town nature of Iligan is manifested in the incomplete or temporary nature of many buildings and in the bamboo and *nipa* slum communities in which live many industrial workers, usually recent migrants from the Visayas. These communities, six in number, occupy over thirteen hectares along the shoreline and river banks. Further evidence of the too rapid development of the city is exhibited by the unpaved street (over two-thirds in terms of length), the absence of sewers away from the main streets, and the high incidence of disease (also, as evidenced by the 1970 Census, fewer people possessed transistor radios—13% compared with the national average of 33%—and electric light—29% compared with 56% for the nation as a whole).

On the other hand, an impressive range of new industrial enterprises provides many employment opportunities. Outstanding is the steel industry, for Iligan boasts both an integrated steel mill and an electric furnace steel plant and rolling mill. A fertiliser plant, cement factory, flour mill, timber and paper pulp plant, and a calcium carbide and ferro-alloy factory also employ many workers. Other manufacturing establishments are concerned with rubber and plastics products, maize, rice, timber, shoes, textiles, food and beverages. To serve these enterprises there has been substantial development of facilities in the fields of banking, hotels, entertainment, and shopping. But, by the standards of the average major provincial centre of the Philippines, facilities offered generally do not compare favourably in terms of either quality or quantity.

As a community also Iligan is relatively unsophisticated compared with most urban centres in the Philippines of similar size. This has its origin in its Mindanao location for, although situated on the northern coast of the island and thus comparatively accessible to Luzon and the Visayas, settlements such as Iligan have naturally been laggard in experiencing and accepting new trends whose origins were in far distant Manila. Community development in the 1960s was given impetus by the arrival of industry but in a direction somewhat different to that of traditional Philippine urban society so that, unlike Dumaguete for example, urban life is not intimate and sedate. The élite of the pre-industrial society may have occupations similar to those of some of the newcomers but often the latter represent more aggressive and ambitious counterparts specially recruited to exploit the boom-town conditions. They exhibit a different residential pattern also; the pre-industrial élite remain orientated to the urban core while the newcomers have their homes in the burgeoning suburbs of Capital Village, San Miguel and Maria Christina. Finally, the industrialist himself may well be less sophisticated and certainly his behaviour often epitomises these dynamic characteristics which can be very much at variance with past standards. The resultant clashes in social outlook are mirrored at the lower levels of society where the migrant workers form an abrasive and unassimilated element of society.

Iligan thus is an urban settlement very much in a state of flux. More such settlements can be expected to emerge in the Philippines in the future, either from planned development or through the metamorphosis of provincial centres. While their urban morphology in the latter case will be traditional, both will face special problems and will require special facilities, like Iligan's training centre for skilled manpower, to meet the challenge posed by the geographical influences which create them.

References

1. A recent study of residential patterns reinforces the conclusion that there is a tendency for above average housing to be concentrated along main highways. D.C. Bennet, 'Aspects of Urban Residential Patterns in Central Luzon', *Philippine Geographical Journal*, Vol. 14 No. 1 (January-March 1970), pp. 20-27.
2. E.L. Ullman, 'Trade Centers and Tributary Areas of the Philippines', *Geographical Review*, Vol. 50 (1960), pp. 203-18.
3. T.G. McGee, *The Southeast Asian City*, G. Bell and Sons, London, 1967.
4. For a dated but still valuable review see W.E. McIntyre, 'The Retail Pattern of Manila', *Geographical Review*, Vol. 45 (1955), pp. 66-80.
5. A. Cutshall, *The Philippines: Nation of Islands*, D. Van Nostrand and Co. Inc., Princeton, N.J., 1965, pp.30-31
6. For a detailed account of *barrio* Trinidad on which this section is based see H. Umehara, 'Socio-Economic Structure of the Rural Philippines: A Case Study of a Hacienda Barrio in Central Luzon', *The Developing Economics*, Vol.7 (1969), pp. 310-31.
7. Source material for this section mainly from: Anon, 'Iligan City Profile', *The Philippine Economy Bulletin*, Vol. 8 (1969), pp. 24-37.

3
The Rural Scene

The events and decisions which shape the life and livelihood of the Filipino may emanate from Manila and the other urban centres but the majority (about 70%) of people affected dwell in the rural areas of the country. Despite its all too common routine and poverty, over half (56.3%) of the employed labour force was engaged in agriculture in 1969.

Isolated individual settlement is not a feature of the rural scene. Home for most inhabitants is the *barrio* of which there are more than 17 000. Each possess some rudimentary central services: some four-fifths have schools catering for at least the lower primary school grades; a weekly market is usually held; and a few *sari sari* stores operate at all times. Social and cultural life operates at a very low level and compares closely with that experienced by the less fortunate members of minor provincial centres, the place to which the *barrio* folk turn for many essential and non-essential aspects of their unsophisticated way of life.

By far the greater number of *barrios* are to be found serving the country's 10 million hectares of cultivated land (Fig. 10), an area equal to that in Japan. Yet such land occupies little more than one quarter of the total land area (Table 3). If not engaged in agriculture, the rural dweller usually gains his living from the sea; *cogon* and other grassland areas are but sparsely settled, for the Filipino is not a pastoralist by nature; elsewhere, a handful of forestry and mining communities share some 16·7 million hectares of forests with the nation's few thousand aborigines.

Geographical Patterns of Land Use

The Philippine rural scene is characterised by a great variety of activities; those pertaining to agriculture present the most complex pattern. Though their local impact is great, mining and

TABLE 3
Land Use in the Philippines, 1968 ('000 hectares)[a]

	Cultivated[b]	Commercial Forest	Non-commercial Forest	Open and Grassland	Marshes	Total
Philippines	10 027	9420	7243	3131	179	30 000[c]
1. Luzon	4469	2890	3427	1229	95	12 110
2. Visayas and Palawan	3250	2001	1519	916	3	7691
3. Mindaneo and Sulu	2308	4529	2297	986	81	10 199
% Distribution	33.4	30.4	24.2	10.4	0.6	100.0

Source: Philippine Forestry Statistics.

[a] All Philippine agricultural statistics should be treated with caution, particularly when making comparisons over time, because of the sometimes unreliable bases of calculation.

[b] Includes 'other' land and may include an element of double counting as a result of multi-cropping practices.

[c] The official land area of the Philippines totals 29 710 000 hectares. But see [b].

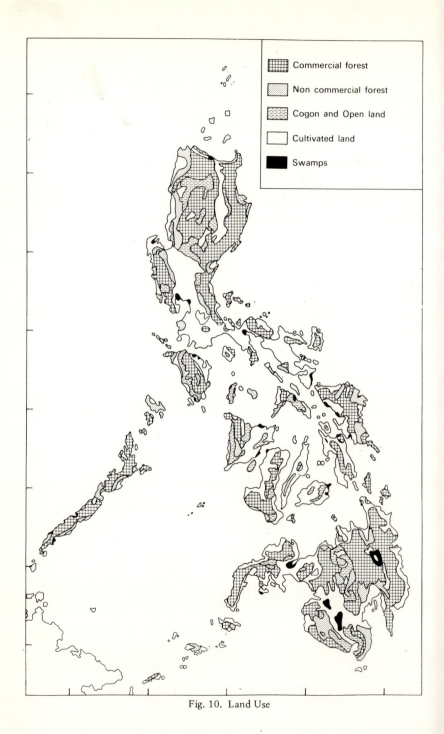

Fig. 10. Land Use

fishing activities generally do not obtrude into the overall rural scene; likewise, the activities of the forestry companies, although often dramatic in certain localities, make but a limited contribution to the nation's overall land use pattern. In contrast, agricultural land use includes not only patterns to be found only in a few specialised locations but also, and more important, patterns which dominate the regional landscapes:

'Northern and Western Negros Island is an agricultural landscape in which sugar cane assumes a dominant status, and in which rice, coconut, bananas, papayas, and mangoes are employed as complements occupying marginal positions in the agricultural landscape. South-east of Manila, coconut plantings dominate the landscape and complementary crops show up in patches, gardens, and homestead plantings. Corn and coconut appear to dominate the landscape in much of Cebu, whereas the root crops present an open landscape in the typhoon-swept Batan Islands. South of Central Luzon the coconut tree has long closed in local landscapes, and as the frequency of coconut planting increases throughout much of the Philippines, its effect on the landscape is increasing. For all the regional specialisations, however, rice, bananas, corn, yams, sweet potatoes, and coconut dominate the cropping systems and the agricultural landscapes of the Philippines.'[1]

In an attempt to simplify the complex pattern of agricultural land use Spencer has enumerated twenty-one crop combinations,[2] but there are doubtless others:

CHIEFLY LOWLAND COMBINATIONS

1. Mixed lowland crops
2. Small scale irrigated rice plus minor crops
3. Large scale irrigated rice
4. Irrigated rice and coconut
5. Coconut and minor crops
6. Plantation ramie and minor crops

CHIEFLY UPLAND COMBINATIONS

7. Caingin upland rice and minor crops
8. Terraced irrigated rice and sweet potatoes
9. Upland rice-coconut-abacá

10. Bananas-vegetables-minor fruits
11. Tobacco and corn
12. Upland rice-coconut-corn
13. Upland rice-cassava-corn
14. Plantation pineapple

COMBINED UPLAND-LOWLAND COMBINATIONS
(found both in upland and lowland)

15. Coconut and corn
16. Corn and minor crops
17. Plantation coconut
18. Plantation abacá
19. Large scale sugar cane
20. Small scale sugar cane and rice (irrigated or upland)
21. Vegetable gardens

Furthermore, any particular combination conceals a much more complex farming pattern:

'The farmer of southern Luzon who plants coconut seedlings and grows upland rice as an intercrop may be classified as engaging in domestic production until the trees begin to bear. Until the growth of the trees makes rice intercropping impractical he will produce in both the domestic and the export sectors. His mature coconut farm will be classifiable in the export sector. The southeast Luzon farmer who produces rice, abacá, coconut, and sweet potatoes as his major crops may sell most of his rice, abacá, and coconut to commercial markets, at which point wholesale dealers dispose of the rice in the domestic market and the abacá and coconut in the export market. This commercial farmer has kept small shares of each of the major crops for home use or local barter trade, saving his sweet potatoes for home use and local barter. But around his homestead he has also produced small volumes of such vegetables as taro, eggplant, and tomatoes, and such fruits as bananas, papaya, lanzon, and avocado. Most of this volume will be eaten at home, but some of it will go to friends, and some will enter the local barter market. His dozen hills of sugarcane provide chewing cane for relaxed evening snacks but do not make him an export producer just because sugarcane is primarily an export crop. His pig and a dozen chickens provide local consumer or barter products at the proper times. The small volume he sells or barters at the local town market helps provide the complementary staples, the regular fish purchases, and the occasional odd purchases.'[3]

TABLE 4
Crop Area Harvested, 1970

	1. Area '000 hectares	%	2. Indices (palay = 100) Area	Value of Output[c]
Palay (rice)	3112.6[a]	34.9	100	100
Maize	2392.2[a]	26.8	75	26
Coconut	1883.9	21.1	61	63[d]
Sugar Cane	366.1	4.0	12	87
Abacá	173.0	2.0	6	5
Tobacco	87.4	1.0	3	6
Other Crops	911.2[b]	10.2	29	105
Total	8926.4	100.0	288	391

Source: Bureau of Agricultural Economics.

[a] 1971 estimates, both marginally smaller than 1970 totals of 3 113 440 hectares and 2 419 600 hectares respectively.

[b] Dominated by root crops (252 400 hectares) and bananas (235 200 hectares), crops which are to be found in almost every *barrio*.

[c] Using 1970 data, index shows relative commercial as opposed to areal importance. The greater significance of sugar cane is clearly revealed, as is the lesser significance of maize.

[d] Includes copra and dessicated coconut but not coconut oil.

Despite this diversity, the rural scene's major components are but six individual agricultural products—rice, maize (corn), coconuts, sugar, abacá, and tobacco[4]—which together occupied 89·8% of the harvested area in 1970 (Table 4). If to any discussion of these major focal points of agricultural activity are added fishing, forestry and mining, there is attained a fully representative sample of rural activity and hence the rural scene. Conspicuous by their absence are livestock (see Chapter Eight) as only pigs (6.5 million in 1970) and carabao (4.4 million) are commonplace and then but rarely in any regional concentration as normally their role is to form an integral part of most cropping combinations. Also conspicuous by its absence is the large scale plantation devoted to tree crops: legislation limits foreign operation of individual holdings to just over one thousand hectares.

Any discussion of the rural scene is simplified by the fact that the basic social and cultural patterns in the rural areas exhibit a high degree of homogeneity, variations in dialect being the only major exception. This statement does not apply to Mindanao where a non-Christian way of life prevails in many parts. This invests the rural scene with a flavour so distinctive that it is discussed in some detail in Chapter Six. The other exotic cultural element in the rural scene—the aborigine—has a much more muted role to play in the shaping of the rural scene. His way of life is primarily of anthropological interest and in consequence is not singled out for special attention (but see Chapter Five). However, the aborigine's one major contribution to the rural scene—*kaingin* agriculture—forms an integral part of the ensuing discussion of the first, and foremost, of the nine basic types of rural activity—rice production.[5]

Rice

Rice, or *palay* as the plant is known in the Philippines, is the main or preferred food of over three-quarters of the population. Rice production, therefore, is widespread in the Philippines (Fig. 11) being grown on no less than 34·9% of all cultivated land (Table 4) and, in volume terms, accounts for about half the output of food crops. Despite the long established popularity of rice the traditional methods employed by Filipino farmers are inefficient: in consequence, yields are low and variable (that for Central Luzon is almost three-quarters as great again as that for the Eastern Visayas) and the product of much backbreaking and unnecessary toil. Thus, the common *tao* often proceeds in his time-honoured way without benefit from fertilisers, insecticides, selected seed, soil conservation and systematic irrigation. For an increasing minority the introduction of HYV rice has brought about a radical change in techniques but it still has far to go before it has a universal impact.

Modern techniques suitable for Philippine conditions have been evolved but as yet only a small number of farmers have been convinced of their efficacy, despite the efforts of the Department of Agricultural Extension. Above average success, however, has been achieved in persuading more progressive rice farmers to utilise the new IR-8 seed, primarily because of the

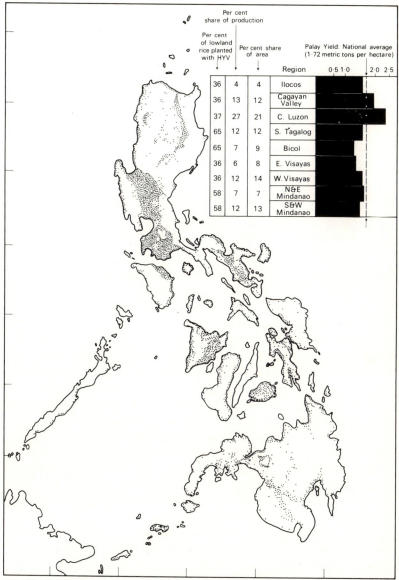

Fig. 11. Palay (dot map is based on latest available—1960—provincial data, regional inset is based on 1971 data). One dot = 1000 hectares

startling increase in yields possible (three times normal or more) with but little change in existing cropping practices. IR-8 rice largely was responsible for the 11·9% increase in output in 1968 (Table 5). In 1970, 50·3% of the national rice area was sown with IR-8 and other high yielding varieties. Through their introduction, the new seeds made possible a 30% increase in the national output by 1971 and the Philippines entered the 1970s as a potential net exporter instead of a net importer of rice.

TABLE 5
Philippine Agricultural Production

	'000 metric tons; crop years					
	1965	1966	1967	1968	1969	1970[a]
Rice	3992	4073	4094	4789	4445	4998
Copra	1471	1485	1418	1542	1643	1588
Maize (shelled)	1313	1380	1435	1619	1733	1787
Sugar (centrifugal)	1557	1402	1560	1595	1596	1885
Abacá (unmanufactured)	134	135	118	103	106	112
Tobacco (raw)[b]	46	58	51	65	57	58
Volume of agricultural production (1955 = 100)	140.0	152.1	155.5	173.9	172.7	182.4

Source: Central Bank Statistical Bulletin, Bureau of Agricultural Economics and Economist Intelligence Unit.

[a] Preliminary.
[b] See note to Table 10.

The low level of management characteristic of *palay* production is in part a reflection of the small scale of production and the operator's often insecure status. The average rice farm is only 3·1 hectares in extent or almost half an hectare smaller than the average for all farms. Of the area only about three-quarters is planted to *palay* in any one year. Furthermore, only about 42% of the rice farms are owner-operated although it should be pointed out that such farms are generally of above-average size, commonly comprising four hectares. In the principal rice growing area (Central Luzon) owner operators are very much in a minority: a typical rice farm here is operated on a tenancy basis and averages only 2·6 hectares in extent.

Rice can be adapted to a relatively wide range of physical conditions and is produced by a great variety of methods. Artifical irrigation schemes are slowly increasing the area suitable for double cropping but as the area involved is less than that of new land opened up for cultivation the total hectarage of rice land is growing at a rate faster than the introduction of irrigation facilities. In the absence of irrigation most rice grown in the Philippines is planted between late April and early June and is harvested some five months later. After the harvest any second crop planted usually consists of maize, vegetables or root crops depending on the farmer's preference and the locality.

Palay cultivation is approached in two distinctive ways in the Philippines which results in the creation of an 'upland' and a 'lowland' scene. The latter is the more widespread, lowland *palay* being characteristic of some four-fifths of the nation's rice land.

The Upland Scene. The cultivation of *palay* in upland areas is commonly associated with shifting or *kaingin* agriculture.[6] However, permanent-field upland rice production is practised also over some half a million hectares or four times the area devoted at any one time to *kaingin*. The upland scene is most commonplace in the southern half of Luzon and Mindanao and to a lesser extent in the Visayas. In the former, as much as one-third of the rice lands are devoted to upland rice.

Kaingin farming involves the clearing of a selected patch of ground which is then farmed for two or three years, by which time declining output causes its temporary abandonment. In selecting a new site the *kainginero* prefers gently sloping land covered by a heavy stand of forest at the base of a hill. However, population pressure coupled with the effects of *kaingin* activity in the past render such sites extremely rare. The modern *kainginero* therefore finds himself limited to steeply sloping hillsides whose depleted soils now only support bamboo, wild grasses or jungle growth. Having determined that the soil character is adequate for *palay* (usually by tasting it to ensure that it is not excessively acidic), the ground cover and any small trees are removed. Larger trees also may be felled but killing by girdling is the simpler and preferred method. Once the cleared material has dried sufficiently it is burned, leaving the field covered with a layer of grey ash and criss-crossed by partly burned branches, tree trunks and stumps. Subsequent cultivation is minimal, the *palay* being planted (or interplanted with other crops, chiefly vegetables) between the refuse with the aid of a planting stick and in a random manner. Some haphazard weeding may occur and, during the ripening season at least, attempts are made to prevent animals damaging the crop. It is not surprising to find that yields are extremely low especially after the first year when soil fertility drops and weed infestation and soil erosion increase rapidly.

In contrast to the haphazard and temporary rural scene created by the *kainginero*, permanent rice cultivation in the upland areas presents a much more familiar agricultural scene by Philippine standards. Permanent-field upland palay cultivation normally is carried out on rolling terrain and involves both ploughing and harrowing of the fields, usually with the aid of some animal power provided by the *carabao*. Seeds are normally broadcast but, in an attempt to get the plants to grow roughly in line, it is sometimes the practice to precede seeding by the use of the *lithao*, a fine toothed wood or bamboo harrow, to cut furrows in which the seeds come to rest.

Despite the relatively careful manner of soil preparation, weeds and the vagaries of local precipitation can be major obstacles to successful cropping. Even given good weather and assiduous weeding yields rarely are high, for continuous cultivation over long periods with little attempt to provide fertiliser and almost no regard for conservation has seriously depleted soils in many areas.

The Lowland Scene. Lowland *palay* not only covers a much greater area than upland palay but its areas of concentration are much more apparent, with the 'rice bowl' of Central Luzon outstanding. Certainly, to the average traveller lowland *palay* is a far more dominant element of the total rural scene. This impression is heightened by the distinctive method of cultivation and its association with the more densely populated rural areas most frequented by and most accessible to the average traveller.

Lowland *palay* cultivation is both labour-intensive and designed for the intensive use of the land. Despite the effort put into its production the cost of the final product is more expensive, both on a volume and a calorie basis, than its principal substitute—maize. As is the case with upland *palay* inefficient techniques play their part but a significant contributory factor is the high cost of land. Good rice land is not abundant in the Philippines. Dyking rather than terracing for the purpose of impounding water is pre-eminent (except in the case of the world-renowned but atypical Ifugao rice terraces). Soils must be a heavy clay to minimise internal drainage and to permit maximum contact between the soil particles and the exceptionally fine roots of the rice plant.

The transplanting system is normally employed, the plants to be used having been grown from the last season's crop in a carefully prepared and tended seedbed. Once the seedlings are sufficiently mature they may be rolled up like a rug and carried to the transplanting field. This is prepared first by flooding to a depth of approximately ten centimetres and then by ploughing. After an interval of a week or so the field is drained and harrowed until the soil is pulverised, level and weed free. After further reflooding, the tiring, time-consuming task of transplanting can begin.

TABLE 6
Lowland Palay: Labour Inputs, 1970 (man days)

	Local Varieties	H.Y.V.	Average
Land Preparation	10	10	10
Pulling and Transplanting	17	17	17
Weeding	7	13	11
Other pre-harvest tasks	9	8	8
Harvesting and Threshing	20	22	21
Total	63	69	67

Source: R. A. Guino and W. H. Meyers, 'The Effect of the New Rice Technology on Farm Employment and Mechanisation', unpublished seminar paper, International Rice Research Institute, Los Baños, Dec. 1971.
Data refers to a sample survey of farms in Central Luzon and Laguna areas only.

Transplanting represents the first of two peak labour demands in lowland *palay* cultivation, the other being harvesting (Table 6). All members of the family (or several families engaged in a co-operative effort) combine to accomplish the task of planting and for each hectare it is necessary to plant roughly 660 000 seedlings. When the crop is ripe, harvesting, hauling, stacking, threshing and winnowing again demand everyone's attention for mechanisation at any stage is very much the exception. In between transplanting and harvesting the rice farmer's tasks are less onerous but nonetheless important. For example, failure to prevent rat infestation can seriously reduce yields: in 1966-7 some 800 000 hectares were damaged by rats and rice worth nearly £40 million was lost.

Mechanisation does play an increasingly significant role in the milling process (and also in the field thanks to the introduction of hand tractors and mechanical weeders) although even today about one fifth of all rice produced in the Philippines is hand pounded for on-farm consumption. As *palay* loses much volume and weight in the milling process it is most economically milled close to the fields. This has led to the growth of large numbers of small and medium-sized mills very widely scattered throughout the *palay*-producing areas of the country. Most common in rural areas is the *kiskisan* (6993 in 1969) which is a small rice mill whose owner is paid in kind at a rate of 5-6% of the rice milled. The *cono* or larger mill (capacity 200-1000 sacks per day) is more commonly regarded as a feature of the urban scene. The four thousand-odd *conos* are especially significant in the Central Plain of Luzon where their individual output may be as much as twenty times that of the average *kiskisan*. Their intake, unlike that of the *kiskisan*, is far from local in origin as the owner operates a fleet of trucks making it possible for him to secure over a wide area *palay* at the most advantageous price. Aside from the economies of scale gained by the *conos* their drying and storage facilities are superior, an aspect of the rice industry as a whole which is substandard and a major factor in its inefficiency.

Maize

Over 20% of the Philippine population consume maize as their principal food staple and over 2 million hectares, or just over one quarter of all cultivated land (Table 4), are set aside for its production annually. Output averages around 1·5 million metric tons annually (Table 5) but is now rising as high-yielding varieties have made their mark in maize cultivation. In 1969 about one quarter of all maize plantings were of varieties capable of providing double the normal yields. The adaptability and flexibility of maize results in its extensive use as a rotation or second layer crop. As a second layer crop maize is most commonly associated with coconuts, while it is usually rotated with *palay*, tobacco, cassava (tapioca) and vegetables. Nevertheless, over large areas of the Philippines, maize is the dominant element in the rural scene.

Introduced to the archipelago from the New World by the Spanish in the sixteenth century, maize has proved to be more nutritious, cheaper to produce, less time-consuming and more tolerant of physical conditions than rice. In consequence maize is being planted or harvested somewhere in the Philippines all through the year. Given optimum climatic conditions (no pronounced rainfall maximum, no dry season, and an absence of typhoons being critical in the Philippines), three crops a year are possible. Most favoured in this respect is the Davao area which, with the island of Cebu, is the focal point for maize production in the country (Fig. 12).

In the major maize producing areas double cropping is most characteristic. The first, and most prolific, stand is generally planted in May or June and reaches maturity in August-October. The second crop, planted as soon as the first is harvested, will be ready in December or January. This planting pattern is so arranged to coincide, where possible, with the two rainfall maximum characteristic of many parts of the Philippines. The planting cycle begins at the time of the previous harvest when seed selection takes place. This is usually done in a haphazard, highly subjective manner. As a result, instead of perpetuating desirable traits the reverse is often the case. The selected ears of maize are stored until the next planting under generally inadequate conditions with the result that both quantity and quality are adversely affected. Furthermore, in the pre-planting period the farmer further diminishes the prospects for good yields by inadequate soil preparation (only one ploughing and harrowing normally being carried out whereas two or three of each are desirable) and by inadequate germination tests, resulting in irregular growth after sowing. Finally, plant maturity is impeded by soil erosion caused by running furrows in the direction of the maximum (often near precipitous) slope of the land; an absence of diligent weed eradication and fertiliser application; and an inability to control the variety of pests and diseases to which maize is subject.

By the time harvesting is ready to commence, the effects of

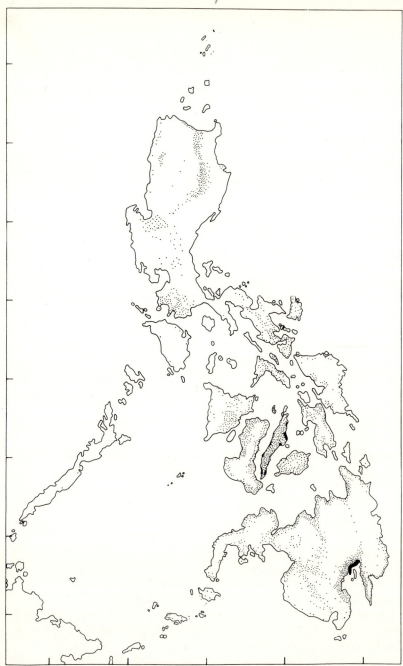

Fig. 12. Maize (based on latest available—1960—provincial data, but from inspection of more recent but incomplete provincial data, no significant variations to the patterns are believed to have occurred). One dot = 1000 hectares

this ill-informed crop management are all too apparent. The stand of maize is of irregular height (rarely exceeding two metres), at varying degrees of maturity, the ears are small and imperfect, and in upland areas soil creep and gulleying are widespread. Unfortunately, this is not the end of the story. Harvesting methods are inefficient: after the hand harvesting and drying of the ears, which as for rice is often a co-operative effort, primitive methods are employed to detach grains from the husk as large dehusking machinery is uncommon. Most involve a variety of primitive devices such as a piece of wood with a metal tooth mounted in a central hole—the tooth dislodges the kernels which fall through the hole to the collection tray; removal by means of the thumb, assisted by the *balibol*, or pointed iron rod, is another common but difficult and tedious method employed. The resultant wastage and grain damage causes yet another reduction in the quality and quantity of the crop. Storage imperfections then cause additional losses principally as a result of the activities of weevils.

Storage losses are most common amongst maize retained for farm use or held in the smaller maize mills located in the *barrios*. The latter, totalling some 3000, closely resemble the *kiskisans*; indeed, 15-20% of these small rice mills process maize. Storage facilities are of a higher standard in the larger mills (488 in 1969) for these are often substantial operations handling up to 10 000 sacks of maize grain daily. They usually are located in provincial centres and engage in the purchasing, grinding, and distributing of maize on a regional basis.

The above imperfections of maize cultivation stem from conditions similar to those indicated for rice, notably a conservative attitude on the part of the farmer, the small scale of production and the general absence of owner operators. The resultant exceptionally low productivity is highlighted by the fact that the average yield for maize is only one third the desirable figure under favourable conditions. With producers concentrating upon increased output by means of expansion onto sub-marginal land (in such areas yields may be barely half the national average) rather than attempting to raise productivity, no dramatic improvements in yields can be expected.

Coconuts

The Philippines produces one quarter of the world's coconuts and the trees from which they are harvested occupy some three-quarters of the area devoted to commercial crops. But the coconut is not solely a plantation crop; it is a source of livelihood for approximately 12 million Filipinos, or almost one third of the present population. As of 1970 an estimated 275 million coconut trees were spread over 1 883 920 hectares or approximately one fifth of the total agricultural land (see Table 4). Annual production normally is of the order of 7·5 million nuts, equivalent to 1·5 million tons of copra. Over 75% output moves overseas in various forms, whose total export value was larger than any other commodity group in 1971 (Table 7)

TABLE 7
Most Valuable Export Commodities[a], 1965-71 (£million)

	1965	1966	1967	1968	1969	1970	1971
Total Exports	320.2	351.3	385.5	353.4	355.8	422.4	n.a.
1. Timber Products	74.8	92.7	93.7	104.4	104.0	103.1	91.5
Logs and Lumber	67.5	85.3	86.4	90.3	92.3	92.1	n.a.
Plywood	7.3	7.4	7.3	8.9	8.0	8.2	n.a.
Veneer	neg.	neg.	neg.	4.8	1 4.5	2.8	n.a.
2. Coconut Products	112.6	111.1	89.2	93.7	57.1	83.6	106.5
Copra	70.8	65.5	53.5	51.3	29.6	35.0	n.a.
Coconut Oil	28.4	31.0	24.3	32.2	20.8	40.5	n.a.
Dessicated Coconut	8.5	7.4	7.1	10.2	6.7	8.1	n.a.
Copra Meal or Cake	4.9	7.2	4.3	neg.	neg.	neg.	n.a.
3. Metalliferous Ores and metal scrap	32.1	50.5	46.4	56.5	75.9	95.5	102.5
Copper Concentrates	19.4	31.1	31.2	37.2	56.5	83.1	77.4
Other Items	11.7	17.4	15.2	29.3	19.4	12.4	25.1
4. Sugar	55.2	555.4	59.0	60.0	62.0	71.1	96.3
5. Canned Pineapple	3.6	3.7	4.3	7.8	7.2	8.9	n.a.
6. Abacá, unmanufactured	10.1	7.8	6.1	4.7	6.0	6.4	n.a.
7. Tobacco, unmanufactured	n.a.	4.6	5.0	7.3	6.7	5.7	n.a.

Source: Central Bank of the Philippines.

[a] Excludes embroideries normally worth about £15 million annually; petroleum products, worth more than £5 million annually since 1967; and animal feeding stuffs normally worth £6.5 million annually.

In value terms, 1970 exports consisted of copra (40%), coconut oil (50%), dessicated coconut (9%), copra cake and meal (1%); in volume terms, 1970 coconut oil exports totalled 338 700 tons compared with 447 400 tons of copra.

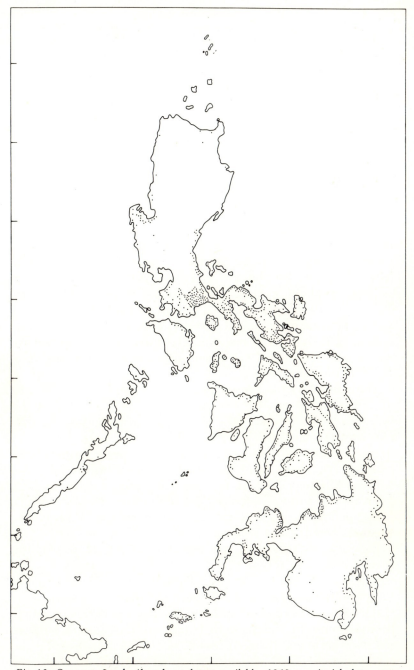

Fig. 13. Coconut Lands (based on latest available—1960—provincial data; more recent regional data is contained in Table 8). One dot = 1000 hectares

The tree is ubiquitous to the south of Manila (Fig. 13) where near ideal conditions for growth are commonplace, i.e. monthly temperatures constantly in the vicinity of 27°C (80°F) and a continuous supply of moisture. Soil moisture can be brackish, though not stagnant; the principal limiting factor to the south of Manila is elevation, temperatures above 600 metres being inadequate for proper growth.

Favoured locales for coconut cultivation are characterised by light, deep, well drained soils with a moderate to high calcareous content. Large scale plantings are thus most often to be found on coastal lowlands or on the rolling lower slopes of recently active volcanic regions. Unfortunately, the tree's susceptibility to damage from high winds significantly hampers production in the Bicol Peninsula and in the Eastern Visayas. In these areas yields may be reduced by up to 25%, a substantial part of which can be attributed to typhoon damage. For example, typhoon damage in Bicol in 1967 was largely responsible for a 15% drop in national output in 1968—to 1·2 million metric tons—and depressed 1969 output also. Further severe typhoons in late 1970 had a similar effect but an impressive recovery occurred in 1971 when output reached a record level of 1·9 million metric tons.

The size of coconut holdings in the Philippines exhibits wide variations. The average size holding is of the order of 4·4 hectares. However, over 30% of coconut farms possess less than two hectares and large holdings (whose size is in excess of fifty hectares) represent only 0·2% of all coconut farms in the country. The large farms are most common in Southern and Western Mindanao, where the average farm size is half as large again as the national average, and also in the Bicol Peninsula. The smallest farms are most common in the Visayas and Northern Mindanao.

Significant changes to the regional pattern of production have occurred in recent years. In 1948 the Southern Tagalog region was the leading centre of production followed by the Eastern Visayas, Bicol and Southern and Western Mindanao, in that order. In 1970 the Southern Tagalog region ranked fourth behind Southern and Western Mindanao, the Eastern Visayas, and Northern and Eastern Mindanao (Table 8). In 1970 the entire island of Mindanao accounted for almost half the country's coconut output whereas in 1958 one third had been

TABLE 8
Coconut Production Statistics, 1970

	Area ('000 hectares)	Trees Total Nos.	Trees Bearing Nos.	Nuts Gathered (million)	Copra	Usage (million)[a] Food	Other
Philippines	1883.9	272.4	215.2	7745.2	7139.6	226.2	379.7
Ilocos	3.4	0.6	0.3	7.4	4.1	2.3	1.0
Cagayan Valley	12.0	1.8	0.7	11.6	Nil	11.6	Nil
Central Luzon	9.0	1.6	1.4	44.3	1.8	40.4	2.1
Southern Tagalog	330.0	56.5	47.4	830.4	644.9	13.1	172.5
Bicol	240.1	30.7	22.1	459.7	389.2	69.7	0.9
Eastern Visayas	387.1	58.7	46.9	2065.9	2065.7	Nil	Nil
Western Visayas	143.1	23.2	19.6	593.0	581.5	11.5	Nil
Northern and Eastern Mindanao	332.7	40.9	29.9	1129.3	942.7	8.3	178.3
Southern and Western Mindanao	425.9	58.5	46.9	2603.6	2509.4	69.4	24.9

Source: Bureau of Agricultural Economics.
[a] Most nuts in the other category are used for dessicated coconut; the nil category also covers negligible quantities, i.e. less than 0.5 million.

derived from the Southern Tagalog region (by 1970 over 750 000 hectares were planted to coconut in Mindanao). The Tagalog region, and adjacent Bicol even more so, have lost ground as a result of typhoon damage and an inability to control the contagious *cadang-cadang* virus disease—both of which are absent from Southern and Western Mindanao. One result of this relocation of production will be a marked upward trend in production in the mid 1970s as the new trees (covering three-quarters of a million hectares) come to maturity.

Coconut is less demanding of labour than any other major crop of the Philippines, requiring annually an average of thirty man days of labour per hectare of established trees. Considerable labour is, however, required to establish a coconut plantation as this involves substantial land clearance, the provision of a nursery, and the transplanting of the most healthy of the sprouted, selected seeds. The young trees also require careful attention preferably involving the use of fertilisers and cover crops or catch crops which, individually or in conjunction, foster growth, reduce soil erosion, improve disease resistance, add nitrogen to the soil and provide a source of income prior to the fruiting of the trees. Once the trees have attained maturity the demand for labour, apart from harvesting and processing, is confined to the removal of weeds and debris, and pest and disease control. The latter is especially important as *cadang-cadang* and the rhinoceros and red coconut beetles can speedily destroy thousands of trees.

As with other of the nation's principal crops, efficient management procedures are confined to a small number of operators. For the majority, mechanisation, fertilisers, pesticides, etc. are rarely employed, with the result that both the volume and quality of output suffers. The coconut is especially prone to such neglect on account of the fact that it requires only limited attention after maturity and is amenable to interplanting with a variety of more demanding crops, thus reducing the farmer's dependence on the coconut as his prime source of livelihood. In the decade 1958–67, for example, the national coconut yield fell by a fifth from 6000 nuts per hectare to 4822 nuts per hectare (typhoon damage reduced yields to 4120 nuts in 1970); in the Bicol region, worst hit by typhoons, yields fell by more than 50%, from 5259 nuts per hectare to 2492 nuts per hectare (1780 nuts in 1970).

Each hectare fully planted to coconut will normally contain 150 trees (the smaller the holding the greater the density and the more haphazard the alignment of the trees). These have to be harvested individually either by climbing the tree, by employing a *halabas* (a long bamboo pole to which a sharp blade is attached), or simply by waiting for the ripe nuts to fall to the ground. The various methods each have their disadvantages, for example, encouraging disease, fruit damage, and harvesting overripe fruit respectively. Ideally, the nut should be harvested when it is twelve months old; earlier or later collection significantly reduces both the quality and quantity of copra. Carefully organised and supervised harvesting procedures are thus vital to efficient production.

Over 90% of the harvest is utilised for copra which is produced on the spot, the meat being extracted from the nut by splitting it by hand with an appropriate instrument, and then dried. Although producing only low quality copra, open kiln drying is the most popular in the Philippines. It utilises the *tapahan*, a framework built over a small fireplace on which the nuts, split in half, are placed meat downwards. Modern kilns which produce the best quality copra are a rarity in the Philippines so the alternative to the *tapahan* is sun drying. On the better-run plantations, sun drying is carefully controlled, the meat being laid on a clean concrete base, turned by shovel each day and covered when rain threatens. Even so, this process only produces medium quality copra. In contrast, the small producer simply sun dries his copra on the edge of the road. It is not surprising, therefore, to find that Philippine copra is generally of poor quality, amongst the lowest in Asia.

The coconut has been called 'The Tree of Life' in view of its ability to provide for a great variety of the local population's needs. Copra, dessicated coconut and coconut oil are the principal commodities derived from the coconut. Large numbers of nuts (200-250 million) are utilised for food and it is possible to produce charcoal from the shell, coir from the fibrous outer husk, timber from the tree trunk and *tuba*, a mildly alcoholic beverage, from the sap of the palm buds. Such is the undeveloped state of coconut production in the Philippines that the potential of these uses has yet to be exploited. For the moment, products of the coconut other than those derived from copra have a strictly local market. Further-

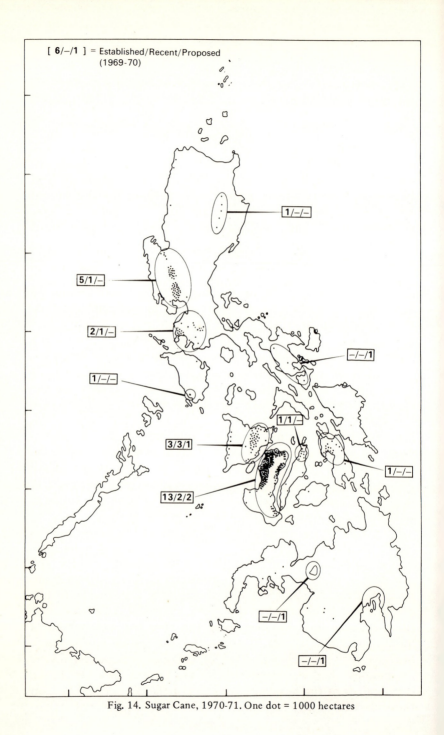

Fig. 14. Sugar Cane, 1970-71. One dot = 1000 hectares

more, in rural areas, where most holdings possess one or more trees, coconut meat may not be used to produce copra; instead that from young coconuts is eaten plain or mixed with milk and sugar, while meat from mature nuts is ground into flour or soaked in water to produce coconut milk, the base for numerous sweets and other foods.

Sugar Cane

Sugar cane country is found over comparatively small areas of the Philippines (Fig. 14) but it is none the less significant. Not only is sugar cane cultivation a dominant element of certain landscapes due to its concentrated distribution but it yields a product whose per hectare value is more than three times that of coconuts and has created production complexes from which the government derives no less than 40% of its annual revenue. Sugar workers and their dependents are estimated to exceed 1 million persons; output normally exceeds 1·5 million metric tons but currently exceeds 2·0 million metric tons thanks to a succession of favourable growing seasons. Two-thirds is exported and in the 1960s earned foreign exchange worth £50-60 million (but £71 million in the peak year of 1970 and then £96 million in 1971), more than any other primary product except coconuts and timber.

The Philippines government exercises considerable control over the sugar industry as a consequence of agreements entered into with the U.S.A. as to the volume of Philippines sugar exports permitted to enter under preferential customs duties. In this, sugar production is unique amongst the agricultural pursuits of the archipelago; with the volume of domestically consumed local sugar representing a calculated surplus over the U.S. export quota, sugar acreages require rigid control. Formalisation of operating procedures goes deeper than this however: 'a unique Philippine system has evolved among world sugar cane producers, a co-operative system involving contract planting and corporate processing in which contracts specify almost every detail of the agricultural operations'.[7]

Sugar cane prefers light, moisture retentive soils, 125-150 centimetres of rain concentrated in the growing season, and temperatures constantly above 30°C. Apart from having growing season temperatures slightly below the optimum, the island of Negros, and in particular its northern part, provides a near-ideal environment. Here, two-thirds of the nation's sugar output is produced on one half of the country's 366 100 hectares of sugar land.

In Negros, there are 2400 sugar planters but the top 10% are by far the most important, and by far the wealthiest. On these large holdings, and on many of the smaller ones, the *haciendero* relies on the efforts of his tenants to produce the sugar cane which is transported by means of narrow gauge railways across the gentle volcanic slopes to the fifteen sugar mills or *centrals*. For his milling and transportation services the *central* owner generally receives 35-40% of the crop value. Of the *haciendero's* share, rather less than half (about one quarter of the crop value) is received by his tenants. In the other sugar producing areas the situation is basically the same although large *haciendas* are less common and less advantageous physical conditions (inferior soil, terrain and rainfall regime) cause yields to fall: in Luzon, for example, one hectare produces only 46 metric tons compared with 63 metric tons in Negros. The importance of sugar to the rural scene outside Negros can be gauged by the number of centrals: Luzon possesses ten, Panay six, Cebu two, and Leyte and Mindoro one each (Fig. 14). Additional centrals are proposed for Bicol, Leyte, Panay, Negros (two) and Mindanao (two).

Little attempt has yet been made to mechanise cane production in the Philippines. The industry is labour intensive, employing at least one worker per hectare, or 10-20 times the input of other countries. In parts of Negros, tractors are used to plough, harrow and furrow the fields after the burning of the refuse left by the previous harvest, but subsequent operations are essentially manual. Labour requirements are highly irregular (Table 9), the massive peak caused by harvesting, and subsequent field preparation and planting, not only resulting in production holdups but also causing under-employment and agrarian unrest in slack times. In Negros, once harvesting commences, 30 000 persons are required for some five months, in addition to the 50 000 persons regularly employed. To meet this labour demand it is necessary to recruit large numbers of workers in the neighbouring islands. Migratory labour is also a feature of the rural scene in Luzon where some 15 000 persons

TABLE 9
Sugar Cane Cultivation: Labour Requirements

	Crop Cycle Month No.[a]	Man Days per hectare	%
Field Preparation	1	10.0	8.0
Planting	1	7.5	6.0
Weeding and Cultivation	2-3	20.0	16.0
Pest Control, other misc. tasks	4-7	2.5	2.0
Harvesting	8-12	85.0	68.0
Total	12	125.0	100.0

Source: R. E. Huke, *Shadows on the Land: An Economic Geography of the Philippines,* Bookmark Inc., Manila, 1963.

[a] Assumed length of twelve months.

travel south from the Ilocos area to help in the sugar harvest on the Central Plain of Luzon.

In the Philippines two harvestings are usually made from a single planting of sugar cane. The first crop will take twelve to eighteen months to mature, the second, or *ratoon*, crop about twelve months. In the third year 'points' from the best canes of the previous year are planted at a density of 30-45 000 per hectare, the points being spaced about twenty-five to thirty-five centimetres apart. Eight to twelve stalks will mature from each of these points. A pointed wooden instrument known as a *topil* may be used to facilitate planting, otherwise the operation is entirely manual. In subsequent months weeding takes place once or twice a fortnight. This too, is largely a manual operation though assisted by a plough or cultivator in the early stages when there is more open ground.

At the same time as weeding, fertiliser has often to be applied tediously, using a tablespoon or ladle made from a small bamboo joint. Various means of assisting cane growth are also employed, notably the turning of the soil away from the young plants to accelerate growth and its later return, known as hilling up, to strengthen the growing cane. Once satisfactory cane growth is ensured the demand for labour slackens markedly for now only irrigation regulation (where practised) and pest control is necessary.

Starting sometime between October and January according to local conditions, another wholly manual operation—harvesting—commences. For the next five months or so the *central* operates at full capacity for twenty-four hours a day in order to consume the cane which has been cut, stripped, trimmed and moved to the railhead points by the *hacienderos'* tenants and the host of temporary assistants. Interruptions to milling are to be avoided at all costs as cut cane rapidly loses its sugar content: a delay in 48 hours in processing can mean a loss of 30%.

With output in Negros averaging over sixty tons per hectare and the average *central* serving some ten thousand hectares, some of which may be fifty kilometres distant, the need for large scale efficiently organised milling is self evident. Cushioned by a guaranteed market in the United States, and a subsidised price on the domestic market, investment in sugar milling (but not necessarily in cropping practices because of abundant labour and the nature of land tenure) has been substantial, and is currently valued in excess of £25 000 000. Consequently, and in contrast to many other areas of Philippine agriculture, methods employed compare favourably with the other sugar producing areas of the world. For example, while occupying only 3% of the harvested area, sugar plantations consume one half of the nation's supply of agricultural fertiliser.

Unfortunately, complacency has gone hand in hand with the financial success of the sugar industry. Schul's comments on conditions in the Victorias Plantation on the island of Negros have general validity:

'Even though the large farm is more efficient as a sugar cane producer, most of the large planters produce far less sugar than they are capable of producing. This fact also results from production being geared to the quota. It is conservatively estimated that at least one third more sugar could be produced in this district if non-restricted sugar production were allowed. The progressive planter is not a necessity since the large planters control more land than they need to attain their quota.
... 'The planter—whether owner or renter—is by nature a promoter of the *status quo*. The planter opposes change for the security of his sugar cane production. However, the propertyless peasant worker cares little if established systems are overturned. Individually he has nothing to lose; he may even gain by a completely new establishment. An agricultural plantation composed largely of landless sugar cane workers is in perpetual peril

since the workers are not reluctant to see constitutional authority weakened or overthrown.

'Under this plantation's system and with the land controlled by a few, there is inefficiency of land use and, correspondingly, inefficiency for sugar cane production. The landless sugar cane worker makes his daily trek into the cane field to insure subsistence. The sugar cane planter, whether owner or renter, has an indifferent attitude toward his sugar cane holding and its employees. If the returns from his hacienda are ample for his well-being, he has little regard as to how efficiently the land is farmed. Furthermore, the planter, who is unusually wealthy for this area, rarely comes into contact with his workers on the hacienda. He resides in one of the urban centers of the Philippines or even overseas. In the Victorias Plantation, 82% of the land is owned by persons living outside the political boundary of the district. These absentee arrangements indicate that few owners care to cope with many problems existing on the haciendas.'[8]

Until the record year of 1970, the Philippines failed repeatedly to meet its quota in the U.S.A. and a decline in export quality was noted also. At the same time economic conditions of the sugar workers have remained sub-standard although adverse publicity has forced employers to formulate new incentives notably a cash bonus scheme which it is hoped will benefit the nation's sugar workers by some £10 million annually. Whatever the success of such measures, the sugar industry will remain out of tune with modern developments. This is most notable where the social conditions of the workers are concerned, for here a feudalistic attitude (or, at best, a paternalistic attitude) is characteristic. In cropping practices, too, progress is not compatible with the industry's standing and the current deterioration in the quality of the cane planted has been accentuated by the exceptional incidence of both droughts and typhoons in the 1960s.

The real problems of sugar production then lie in the field not in the factory. Here, soil improvement, soil conservation, moisture control, increased mechanisation, cane breeding, pest and disease control, and land reform would result not only in a much more efficient and productive sugar land but also a much more tension free and progressive rural scene.

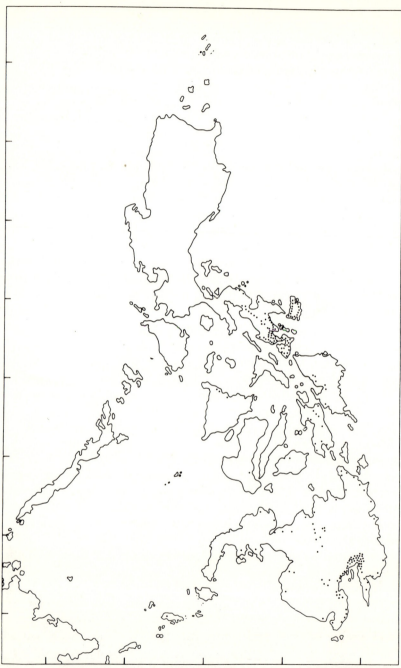

Fig. 15. Abacà (based on latest available—1960—provincial data; subsequent changes have involved overall reduction in hectarage but no substantial change in distribution). One dot = 1000 hectares

Abacá

Most of the nation's 173 000 hectares of abacá are to be found in the vicinity of Davao and the Bicol Peninsula (Fig. 15). Production generally is from small-holdings but a few plantations survive in the Davao area. Yields in the Davao area can be about one quarter greater than in the rest of the Philippines, primarily due to superior physical conditions and greater mechanisation. However, in recent years failure to control pests and diseases not only has forced farmers northward from the original centre of production around Davao but also has reduced the overall importance of the region as a whole.

The Philippines has a virtual world monopoly of abacá fibre production and its principal end-use, rope, is in great demand. Yet current exports are well below pre-war levels, a prime cause being the fact that yields are well below pre-war levels. This state of affairs stems from a variety of causes. Unskilled exploitation in the post-war years resulted in the subdivision of most plantations and a lack of interest in plant care and experimental work. More important, an absence of pest and disease control spelt ruination for large areas, mosaic disease and the *pague pague*, or slug caterpillar, being especially devastating in their effects. With world markets being captured by synthetic fibres too, the abacá industry has been beset with problems. The result has been a lack of incentive to improve management techniques and the conversion of many plantations to other crops resulting in an average annual decline in output of 7% in the mid 1960s. Nevertheless, although abacá is no longer the force it was as an export earner, its foreign currency earnings in 1970 exceeded £6 million. Prospects for domestic consumption have been enhanced by abacá's new use as a raw material for paper pulp: this led to the first increase in output of any consequence for seven years in 1970. In that year domestic consumption totalled about 25 000 metric tons compared with exports of nearly 60 000 metric tons.

Abacá land in the vicinity of Davao is located either on the coastal lowlands or on the adjacent gently rolling hill areas. In the Bicol Peninsula, abacá plantings are generally located on the margins of the lowlands and on the lower slopes of the numerous volcanoes. The optimum conditions existing in these areas are primarily the product of physical factors but a good supply of labour and access to cheap transport also are significant. The abacá plant, a smaller, more slender relative of the banana, flourishes in the humid climate of Davao and Bicol with its year-round rainfall. The local soils too are to its liking being well drained loams rich in organic material, those derived from recent basic volcanic material being ideal. Davao possesses an additional advantage in that only negligible wind damage is caused by typhoons: in the Bicol Peninsula, interplanting of coconuts and similar species is necessary to act as a windbreak.

Abacá cultivation requires a combination of skill and hard work (Table 10). New plantings are made from suckers from the

TABLE 10
Abacá Cultivation: Labour Requirements (man days)

		Bicol and Visayas	Mindanao (Davao)
A.	Development (First Two Years)		
	Land Preparation	14	14
	Planting, inc. shade trees and cover crop (Mindanao only)	17	20
	Weeding, underbrushing and cutting of dried leaves	58	58
	Fertiliser application and pest/disease control	23	36
B.	Operation (Succeeding Years)		
	Underbrushing and cutting of dried leaves	20	20
	Fertiliser application and pest/disease control	12	18
	Harvesting	54	107
	Stripping and bundling	58	58

Source: Abacá and Other Fibres Development Board.

base of a mature plant or, where the absence of interplanting reduces shade, by means of corms or root stocks. Maintenance of soil fertility is vital but a continuing supply of organic material is achieved only by means of leaves and other waste from harvesting which cover the ground between the plants (a device which also curtails moisture evaporation). Leguminous trees such as *anee* and *dapdap* may be employed to offset the absence of nitrogenous fertiliser, to provide shade and serve as a windbreak. Alternatively, a cover crop of *mungo* beans may be employed: this has no shade or windbreak effect but adds nitrogen to the soil, controls weeds and provides the smallholder with a source of food.

Once the plant is established only weed, pest and disease control take up much labour prior to harvesting. Abacá is ready to harvest eighteen months after planting and can then be cut again every three months. Most varieties can be cut in this way for at least twelve years without any appreciable diminution of yield. Harvesting and subsequent stripping are usually accomplished by teams of two people. A scythe-shaped knife on a long stick removes the fronds; a broad square-tipped *bolo* serves to fell the stalk. A *loknit*, a needle-like blade (or sharpened piece of hardwood), is then employed to peel the leaf petioles from the stalk. Next, the most laborious task of all involves the extraction of the fibre from the pulp. Generally, to achieve this each strip is pulled by hand between a polished hardwood block and a knife blade held firmly against the block by a bamboo spring. If the knife is serrated it is easier to strip the fibre but the larger the teeth and the farther apart they are spaced the poorer the quality of fibre produced. In Davao, where the optimum conditions produce larger plants, the increased difficulty of hand stripping has resulted in the wide acceptance of the spindle stripper or *hagotan*. Pulling is accomplished by mechanical power, the strip being attached to a rotating cone-shaped spindle. The resulting fibre is finer in texture and whiter and more thoroughly cleaned than hand-stripped fibre. Output, at ninety kilos per man day, is ten times that for hand stripping. On some of the larger Davao plantations decorticating machines have been employed but this has not proved entirely satisfactory as the method produces fibre with a low tensile strength.

Having been separated, the fibres are crudely sorted for colour and quality and then sun dried. Once dry they are tied into loose bundles for sale to the dealers. Prior to export, government licensed brokers sort, classify and bale the product.

Tobacco

In two localities in the Philippines—the Ilocos coastal strip and the Cagayan Valley—tobacco is the dominant element of the rural scene (Fig. 16). To the casual observer, this might not

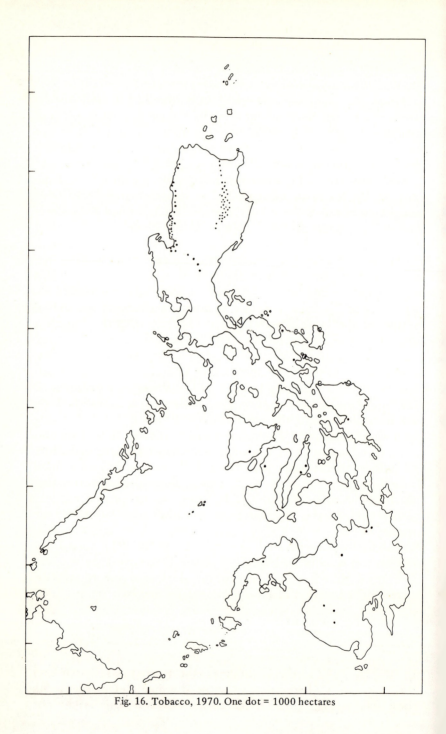

Fig. 16. Tobacco, 1970. One dot = 1000 hectares

appear to be the case as tobacco usually is a second crop, planted on land which already has yielded maize or rice. But, although the success of the cereal crop means a plentiful supply of cheap food, the prosperity of both areas depends on tobacco—Virginia leaf in the Ilocos coastal strip and Native leaf in the Cagayan Valley. Together, the two locations account for some two-thirds of the nation's tobacco lands.

Over 100 000 hectares of tobacco land existed in the peak year of 1962 and the two principal varieties then were almost equal in area. By 1970, however, less than 90 000 hectares were under tobacco (Table 11), with a further, temporary, drop in area expected for 1971 following severe flooding in the Cagayan Valley. It is Virginia leaf which has suffered a cutback; initially, because of overproduction (output jumped from 10 097 metric tons in 1955 to 34 232 metric tons in 1960 in response to the growing taste of Filipinos for American style cigarettes containing a blend of local and imported Virginia leaf) and, more recently, by falling demand as the rising cost of imported leaf (especially following the 1970 *peso* devaluation) caused an enforced switch in taste back to cigarettes containing Native leaf. Thus, by 1970 Virginia leaf accounted only for 39·4% of the Philippine tobacco lands and its share of production volume had dropped from 53·4% in 1960 to 36·0% in 1970. Even more striking have been movements in unit values of production: between 1960 and 1970 that of Virginia leaf rose by only an insignificant 26% whereas that for Native leaf increased almost five times, an impressive performance even after allowing for the depreciation of the *peso* in the sixties.

Throughout the Philippines it is estimated that there are nearly 110 000 individual producers of tobacco supplying the 56 tobacco products factories that existed in 1970. The large number of suppliers reflects the typically small scale of production. Tenant operated holdings in the Cagayan Valley range between two and three hectares in size. However, the area planted to tobacco is dictated by the small number of corporations who own the land so the average area cropped is only 1·5 hectares. The comparable figure for the predominantly owner operated Ilocos coastal strip is in the vicinity of one hectare. Elsewhere in the archipelago even smaller tobacco plantings are to be found.

Rainfall is the dominant influence on tobacco production,

TABLE 11
Tobacco Statistics, 1960-70

Year	Area ('000 hectares)			Production ('000 metric tons)			Value of Output (*pesos* per kg)		
	Virginia	Native	Total	Virginia	Native	Total	Virginia	Native	Total
1960	51.7	44.1	95.8	34.2	29.8	64.0	1.76	0.44	1.15
1965	28.8	47.3	76.1	17.2	28.6	45.8	1.70	0.55	0.98
1966	25.4	60.3	85.7	14.9	43.3	56.2	1.47	0.64	0.85
1967	24.4	57.7	82.5	14.7	36.4	51.1	1.67	0.82	1.06
1968	28.7	64.9	93.6	17.4	47.5	64.9	1.76	0.99	1.20
1969	32.3	57.1	89.4	19.9	36.9	56.8	2.03	1.74	1.84
1970	33.4	54.0	87.4	22.0	39.2	61.2	2.22	2.05	2.11

Source: Bureau of Agricultural Economics.

the plant being able to thrive throughout the Philippines even during the coolest part of the year. Areas with a monsoonal-type climate are best suited for tobacco. A pronounced monsoonal regime is a disadvantage in that it results in strong, coarse, inferior quality leaf. The most desirable climatic conditions are to be found in the Cagayan Valley which experiences a short dry season with a mild rainfall maximum in the autumn and winter months. In this area the production of high quality Native leaf tobacco is enhanced by the presence of fertile alluvial soils with a near-perfect combination of trace minerals. The result is a leaf, neither too gummy nor too tough, which has a high reputation as a cigar filler; it is also used in local-style cigarettes. Virginia tobacco is produced in the Ilocos region. Here the soil, which exhibits limited nitrogen availability but significant amounts of phosphorus and potassium, gives a tobacco of pleasant aroma and fine flavour acceptable for the production of American-style aromatic cigarettes.

Tobacco is grown from seed, being sown in specially prepared beds in either August-November (Virginia leaf) or September-October (Native leaf). Transplanting takes place six to nine weeks later, a week or ten days before this event the seedlings having been hardened by removing their protective shade cover. Despite the introduction of mechanisation, the seedlings are generally hand planted into fields which have been thoroughly ploughed and harrowed. Two or three periods of cultivation, the first by hoeing, the later ones by ploughing, are then necessary to strengthen the young plants by means of the removal of weeds, the loosening of the soil and the uncovering of buried plants. Once growth is established it is necessary to facilitate leaf growth by removing the plant's lower leaves, any flowers, and the suckers which develop as a result of flower removal. During this time attention is also paid to the control of pests and diseases, the most serious of which are the diseases known as tobacco mosaic, frog eye and damping off. Control is largely a matter of prompt discovery and the application of generally simple and well tried preventive methods.

Harvesting generally begins seventy-five days after transplanting and involves hand picking which is done from four to eight times at five to ten day intervals in the season. As only one to three leaves are picked from each plant at any one time all leaves are enabled to reach maturity. Then follows the important process of curing. Native leaf tobacco is sun dried for a day or two and then the leaves are hung for twenty to thirty-five days on the racks of the curing shed which is usually a small hut often flimsily built from bamboo with *cogon* roofing. Subsequent fermentation is carried out by piling the partially dried leaves in stacks. The heat thus generated improves the colour, texture and flavour of the tobacco. Finally, ageing from one to five years further improves the taste and aroma for which Philippine cigar tobacco is internationally known. In contrast, curing of Virginia leaf involves the use of flue curing barns. Usually five metres square and five metres high, some 15 000 of these barns are to be found scattered throughout the Ilocos coast. The barn, airtight and fully filled, is heated through flue pipes to encourage proper yellowing and drying, a process which takes from seventy to a hundred hours. All the above processess are demanding of labour. One hectare of Native leaf tobacco consumes some eighty-three man days plus a further nineteen animal days. For Virginia leaf requirements are almost double.

Virginia tobacco producers in the 1960s were the recipients of a substantial government subsidy—approaching £10 million per annum—aimed at stabilising prices. The inefficiencies of the administering body, the Philippine Virginia Tobacco Association, coupled with changing demand led in 1970 to the adoption of a private enterprise approach focussed initially on the Phil.-Asia Trading Corporation. Its impact was sufficiently favourable to encourage a broadening of private enterprise participation in Virginia leaf marketing making possible a reduction in the government subsidy to below £1 million per annum.

Government assistance to the tobacco industry also has contributed to regional development. Yearly, a sum in excess of £50 000 has been set aside for the construction of feeder roads in regions where the culture of shade grown Sumatra type (cigar wrapper) leaf has been found suitable, notably in the province of La Union. Some 100 kilometres of feeder roads have been constructed or improved in this manner. Such developments suggest that the potential of the industry has yet to be realised. But this hinges as much on the development of existing resources as on new ones. In this context, improved production efficiency resulting from farmer education and the elimination

TABLE 12

Fisheries Statistics, 1965-9 ('000 metric tons)

	Production				Nutritional Fish Allowance (per caput)[b]	Production Deficiency (%)
	Commercial Vessels	Fishponds	Other[a]	Total		
1965	300.1	63.2	303.9	667.2	991.7	32.7
1966	314.9	63.7	326.7	705.3	1026.4	31.3
1967	330.9	63.9	351.2	746.0	1062.6	29.8
1968	406.8	86.7	444.2	937.7	1309.7	28.4
1969	368.7	94.6	477.5	940.8	1356.3	30.6

Source: Philippine Fisheries Commission.

[a] Municipal fisheries and sustenance fishing, i.e. with vessels of less than three tons.

[b] The per caput allowance subsequent to 1967 was raised by 5.84 kilograms to 36.5 kilograms following a reappraisal by the Food and Nutrition Research Centre.

of marketing malpractices are as vital as measures to stabilise prices. Only then will the rural scene in the tobacco lands be an example to other parts of the country.

Fishing

Fish is second only to rice as the basic food of the Filipino, supplying more than two-thirds of the protein produced in the country. Nevertheless, annual production of fish, although at about the 100 000 metric tons mark, has yet to meet the requirements of the fast growing population. In 1969 total fish output amounted only to 69·4% of the fish nutritional needs of the people (Table 12).

At first sight the inability of the Philippines to meet its fish needs is surprising for the coastline is over 17 000 kilometres in length and for every hectare of land in the archipelago there are over six hectares of inshore waters (Fig. 17). In addition to these 1·9 million square kilometres, fishing is possible also over some 1·5 million hectares of freshwater and brackish swamps and lakes, and over 160 000 hectares of existing fishponds (potential fishpond sites total a further 500 000 hectares). Finally, flooded rice fields, local streams and foreshores yield a not inconsiderable haul of marine and freshwater life.

Unfortunately, the fishing grounds of the Philippines are not especially productive; a night's fishing, for example, may only reward each participant with a few kilos of assorted fish. (The absence of an extensive continental shelf and a predominance of warm currents with little upwelling of mineral-rich sub-surface water result in a great variety of fish, each of but limited volume). The low quality of the natural resources is only partly to blame as inadequate techniques are commonplace. In the case of freshwater areas, considerable depletion due to over intensive fishing is an additional factor. Some four-fifths of the marine catch is derived from the Sulu and Visayan Seas (Fig. 17); Manila Bay (less than 10%) is the only other major fishing ground. Navotas, on the shore of the latter, is the country's leading fishing harbour handling almost two-thirds of the 1969 commercial catch (if other Manila Bay fishing ports are included, notably Malabon and Manila [North Harbour] the amount handled rises to over 70%).

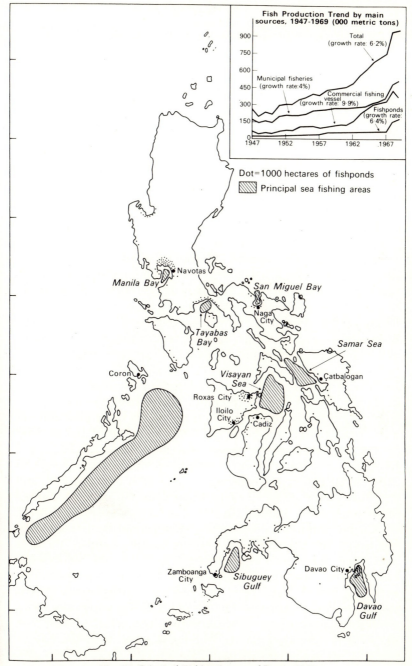

Fig. 17. Principal Fishing Areas and Ports, 1969

Despite the difficulties it faces, the Philippine fishing industry is making progress. In the five years ending in 1969, production increased by 41·1% to reach 940 792 metric tons. The expansion of commercial fishing did not keep pace with that of municipal and sustenance fishing despite the 27% increase in registered tonnage which took place in this period. Municipal and sustenance fisheries in 1969 produced 477 500 metric tons, or 29% more than the commercial fisheries whose development was retarded by operating inefficiencies. Production from the third major source—fishponds—largely was stagnant for the 36% increase recorded in 1968 primarily was the product of a statistical adjustment to allow for previous underreporting of the output of private ponds.

Relatively little fishing takes place outside the territorial waters of the Philippines. Motorisation of the small craft employed by the vast majority of commercial fishermen has taken place at a rapid rate to reach 92% in 1969 but generally boats are not equipped to stay at sea for more than twenty four hours at one time. Inshore fishing thus predominates: of the commercial species, scad (28% of the catch) and slipmouths (13%) are most important, followed by nemipterid and sardine (6% each) and anchovies (5%). Pelagic species account for over half the supply of all fish produced in the Philippines.

Numerous techniques are employed for offshore fishing, the bagnet (796 units in 1969), the otter trawl (667 units) and the purse seine (223 units) being the principal means of fish capture on any significant scale. Their relative importance can be gauged from the contribution of various techniques to the 1969 commercial fish catch: otter trawl (36·4%), bagnet (31·2%), and the purse seine (21·7%). The bagnet, or *basnig*, is a mobile impounding net, cubical in form, which is suspended from detachable bamboo horns employed to spread the net. On dark nights fish are attracted by a strong carbide lamp to the area above the net, which is then raised, thereby impounding the catch. Otter trawling involves a net in the form of a conical bag with the mouth kept open by various devices and which is towed, trailed or trawled on the sea bed. Trawling operations are restricted by rocks and coral formations, which characterise much of the fishing grounds. The purse seine, with the powerblock (100 units), has superseded the round haul seine, or *lawag*. The latter requires a large fishing outfit of sixty to eighty men and several vessels. The net is spread against the current and the fish are attracted to it by lights stationed on small boats strategically placed around the rim of the net. This method catches more fish than any other but involves a disproportionate amount of manpower and incurs high maintenance costs.

The above mentioned techniques are employed both by the principal fishing communities and by the large commercial operators (those possessing boats of three tons or more). The latter, who employ some 45 000 persons, are based at the nation's principal ports, notably Navotas, and often are linked to fish processing concerns. Increasingly, the commercial fleets are being serviced by fish carriers (225 units in 1969) making possible prolonged visits to increasingly distant fishing grounds. For the majority of the nation's fishermen, however, methods of fish capture revolve around the handline, single or multiple; the fish corral, a bamboo constructed semi-permanent trap;[9] the fish pot; and similar simple devices. It is estimated that about a quarter of a million fishing families in 6000 marine towns and *barrios* base their livelihood on such devices. The number of active fishermen (excluding some 150 000 persons employed in fish ponds) is estimated at half a million, who have at their disposal about 5000 fish corrals and some 172 500 fishing craft.

Fisheries in inland areas become significant features of the rural scene on the northern margins of Manila Bay and on the island of Panay (Fig. 17). Here, the brackish water of the swamplands has been impounded in many areas to create a series of artificial fishponds whose per-hectare cash return exceeds even that for sugar land. The milkfish, or *bangos*, is the chief inhabitant of the fishponds. Its development is carefully supervised in the more highly developed areas; this involves transfer of the fish through a series of specially constructed and regularly fertilised nursery ponds (*pabiayan*) and detention ponds (*impitan*) as it increases in size. The result is continuous production throughout the year. More primitive fishpond culture is practised in the Visayas and Mindanao; there, output levels are depressed by a high mortality rate and the undesirable mixing of species.

Inland freshwater fisheries are variable both in their nature and in terms of species caught. Most important are Laguna de

Bay in Luzon where the catching of *kanduli* is a major occupation of the lakeshore inhabitants; the freshwater swamps of the Central Plain of Luzon where large catches of mudfish are taken; and the goby fry and mullet which can be secured from streams and lakes throughout the Philippines when swarming takes place.

The larger scale operators have access to the fresh fish markets of the major urban centres and to fish processing plants. Fishing barrio catches are usually handled by fish merchants who salt, dry or smoke the fish. Smaller fish may be converted into paste. Time-honoured, crude methods are generally employed despite government encouragement to raise standards. Fresh fish appears on the rural dweller's menu only if he lives within a few miles of the sea. For the majority, fish eaten takes the form of *tuyo* (salted, dried herring type fish), *daeng* (salted, dried mackerel type fish), *bagoong* paste (salted anchovy or shrimp), *patis* (fish sauce) and *tinapa* (smoked salted *bangos*). The rural dweller may not live in luxury but his fish diet is rarely sparse and certainly is varied.

Forestry

Nearly one third of the Philippines is classified as commercial forestland (Table 13). Virtually all of the 9.4 million hectares involved belong to the State, which derives considerable revenue from the forestry industry. Some 3000 species present in the Philippines will produce marketable timber, but currently less than sixty species find their way onto the market in significant quantities. Existing reserves are estimated conservatively at 600 million cubic metres. Three-quarters of output is derived from the dipterocarp (*lauan*) family, which yields timber known to the trade as Philippine mahogany. Dipterocarps often occur in relatively pure stands and can yield twenty-four cubic metres per hectare. National output rose by three-quarters in the period 1964/65-1969/70 to reach 11.0 million cubic metres. In terms of export earnings timber products now surpass the traditional leader, coconut products, being worth about £100 million annually (Table 7). In 1950 timber accounted for only 2% of the value of exports; in 1970 it accounted for 29%, despite depressed conditions in the wood processing sector

TABLE 13
Philippine Lumber and Logging Industry, 1970

	Nos.	Daily Capacity[a]	Capital Investment (£ million)	Production[a]	Exports[a]	Exports as % of Production
Sawmills	374	0.018	17.2	11.0	9.6	86.7
Plywood Factories	24	0.133	11.4	21.3	9.1	42.8
Veneer Factories	23	0.260	8.9	18.4	18.4	100.0

Source: Fookien Times Yearbook, 1971

[a] Millions of square metres except sawmills, which is million cubic metres

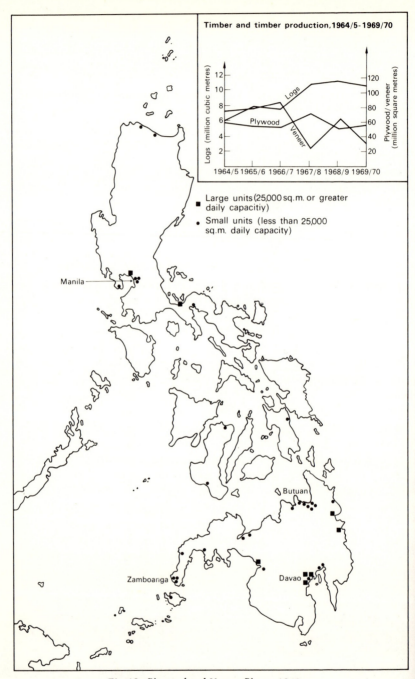

Fig. 18. Plywood and Veneer Plants, 1968

caused ironically by competition from processors in the traditional centres of demand for Philippine logs—Japan, S. Korea and Taiwan.

Forestry operations disturb only a small part of the Philippines forests at any one time, although their long term effect is to modify the natural vegetation complex. However, the use of modern sawmilling equipment and heavy machinery for logging is commonplace in the Philippines so, where practised, forestry operations result in dramatic modifications to the forest scene. An exception is the activities of the small scale operators, often working illegally, who usually cater for strictly local demand. Their output is supplemented by the exploitation of bamboo groves, mangrove swamps and other areas of non-commercial forest (such forestland, together with forested areas too inaccessible or too ravaged by the *kainginero*, approaches 4 million hectares in area). Its products make a significant contribution to fuel supplies, and to house and artifact construction, especially in the rural areas.

Forestry operations are focussed upon the 400 or so processing plants scattered throughout the country, being located with respect to either the principal commercial forests (Fig. 18)or the centres of consumption, notably Manila. In 1970 the country possessed 374 sawmills, 24 plywood mills and 23 veneer mills, including 16 integrated mills producing both plywood and veneer. Capital investment involved totalled £37·5 million (Table 13). Plant ranges in size from family operated sawmills to large scale integrated operations employing over 1000 persons. The latter, like the *centrals* of the sugarlands, are not present in large numbers but they make a major impact on the local rural scene. They may warrant designation as a political entity (a *barrio* or perhaps even a municipality); such giants will be in effect company towns incorporating hospitals, schools, etc. Such operations are most common in Mindanao: Luzon may possess more sawmills (195 to Mindanao's 138) but the capacity of Mindanao's forestry operations overall is nearly double that of Luzon, and includes over three-quarters of the nation's plywood and veneer capacity (Fig. 18).

Ease of access is a major factor controlling the location of logging operations. Logging in coastal lowlands and adjacent foothills is a once only operation for the land is then opened up to agricultural settlement. Inland, as slopes increase and distances from the coastal based processing plant increase, only terrain amenable to the construction of heavy duty logging roads can be considered. Most favourable for exploitation are areas traversed by a waterway sufficient in size to serve as a mode of transport for logs and possessing at its mouth a site suitable for the establishment of large scale processing facilities.

Remoteness is commonplace: for example, virtually the only economic activity on the rugged exposed Pacific coast of Mindanao is a timber processing complex at Bislig Bay giving employment to 1700 persons and supporting a community of some 30 000 persons. Here, using timber from concessions totalling 275 000 hectares, some 500 000 cubic metres of timber are extracted annually. This timber is moved along some 400 kilometres of roads to a deepwater port which handles some 150 ocean going vessels annually. At the port is to be found one of the most modern automated plywood mills in the Far East, together with a sawmill, a veneer mill and, last but by no means least, an integrated pulp and paper mill, the first in South-East Asia.

Conditions especially suitable for logging are most commonly found in Mindanao which possesses 55% of the nation's timber reserves (Fig. 10). Here the absence of any large scale, long-standing settlement, too, has preserved much of the indigenous timber. Most islands of the Visayas are too small or too denuded of timber to warrant large scale logging although the provinces of Negros Occidental, Samar and Antique are not unattractive to the logger. Mindoro, too, possesses a considerable forest cover but here, as in Palawan, costs of development in these pioneer regions are high. The fact that the principal species (*molave, narra, ipil,* etc.) are not of the dipterocarp family has retarded development also. These species, though noted for their beauty and durability, yield relatively little timber per hectare.

In Luzon, which possesses 20% of the nation's timber reserves, the forests of Quezon and Mountain Provinces remain large enough to attract the forestry industry, although for reasons different to those operating elsewhere. In Quezon Province, the very rough topography has not deterred operators who have the incentive of the large market for construction timber and plywood in nearby Manila. Tramline, highlead and skyline logging have to be employed but the additional costs of these techniques are offset by the ease of access to the lucrative Manila market. In the quarter of a million hectares of commercial forest located in Mountain Province, a change of species caused by the elevation further modifies the nature of forestry operations. Only Benguet pine is logged in any quantity and, because of priority given to watershed conservation, such timber as is produced is restricted to consumption exclusively within the Province, the local mining companies in particular being major customers. Here, as in Quezon Province, large scale processing plant is not a feature of the forest scene for the rugged terrain restricts the opportunities for co-ordination of logging operations.

Mining

Mining activities, like those associated with forestry, are limited in areal extent but are concentrated in impact. In 1970 copper exports almost displaced coconut products as the country's second-most valuable export commodity while the output of all mining activities was valued at nearly £190 million. The only operation that yields metallic products is the mining and milling of gold and mercury ores which results in gold bullion and crude metallic mercury. Outside these activities, mining operations are focussed on the selective extraction of raw ores and minerals which subsequently are exported unprocessed or as recovered concentrates. In the case of base metals, ores may be milled to reduce freightage and to meet smelter specifications. Excluded from the following discussion are cement and other construction materials; in 1968, they accounted for almost one third of the value of Philippine mineral output with cement, from twelve plants, pre-eminent.

Copper, iron, and gold, in that order, are the minerals attracting the greatest attention at present (Table 14), with nickel the leading contender for the title of the 'mineral of the future'. Employment in the mining industry totals some 34 000 persons. A general absence of extensive and proven mineral rich deposits means that the impact of mining on the overall rural scene is sporadic. Its local impact, however, is reinforced by the presence of abandoned mines and prospecting activities, both of which are a not inconsiderable factor in a country such as the

TABLE 14
Mineral production, 1967-71

	Unit of Volume	Volume				Value (£ million)			1971 (est.)
		1967	1968	1969	1970	1967	1968	1969	1970	
Precious Metals										
Gold	'000 oz.	500	527	571	603	11.1	11.5	12.7	14.0	16.2
Silver	'000 oz.	1396	1575	1534	1700	0.8	1.4	1.0	1.7	n.a.
Base Metals										
Copper	'000 metric tons	91	110	128	160	37.3	50.3	70.3	118.8	132.3
Chromite Ore	'000 metric tons	420	439	468	566	3.3	3.3	3.7	6.8	5.7
Iron Ore	'000 metric tons	1506	1353	1560	1870	6.0	5.0	5.7	8.9	11.0
Mercury	Flask of 76 lbs.	2544	3543	3478	4648	0.5	0.8	0.7	1.1	n.a.
Non-Metallic										
Coal	'000 metric tons	70	32	53	42	0.2	0.1	0.1	0.1	n.a.

Source: Central Bank of the Philippines, Bureau of Mines, Fookien Times Yearbook, and Economist Intelligence Unit.

Philippines which is endowed with a diversity of base rocks of known or potential mineral content. As with forestry, associated transportation facilities further contribute to the creation of a markedly individualistic rural scene.

Some two-thirds of Philippine gold comes from the Baguio region of Mountain Province (Fig. 19). The mineralised zone is large by Philippine standards, covering over 10 000 hectares. Shaft mining is necessary to extract the gold and the other minerals—copper, lead, zinc, tellurium, etc.—with which it is associated. Gold is found in various other localities, notably in Camarines Norte, Masbate and Surigao, but abandoned workings are more conspicious than active ones, for a substantial proportion—about one third—is now obtained as a by-product of copper mining.

Iron ore generally is found in lateritic form so opencast mining is typical. Such techniques have a much more drastic impact on the rural scene, but in only two areas of the Philippines, Camarines Norte (the Larap Peninsula) and Ilocos Sur, are any major operations well established. Output from these deposits accounts for over two-thirds of the nation's iron ore production (which totalled 1·87 million metric tons in 1970 and was valued at £8·9 million). Shortly, their dominance will be challenged by operations in the Surigao area, where extensive prospecting of iron-nickel deposits covering 20 000 hectares has confirmed the feasibility of large scale exploitation, and also, on a smaller scale, from the nearby magnetic iron sand project near Tacloban, Leyte, where a 1000 metric tons per day capacity unit commenced operations in 1970.

Copper mining makes by far the most important contribution to current mineral output. Some three-quarters of base metals production consists of copper whose export value in 1971 totalled £77.4 million (in 1968 the comparable figure was only £37·2 million). Although more widely distributed than most other Philippine minerals, attention has been focussed on the principal deposits where large scale operations reduce extraction costs. Consequently, three locations are pre-eminent amongst the fourteen currently operative: Toledo (Cebu), Lepanto (Mountain Province), and Tapian in the centre of the island of Marinduque. In 1970 Toledo (the largest open pit copper mine in the Far East) accounted for 27% of Philippine output, compared with Tapian (21%) and Lepanto (16%).

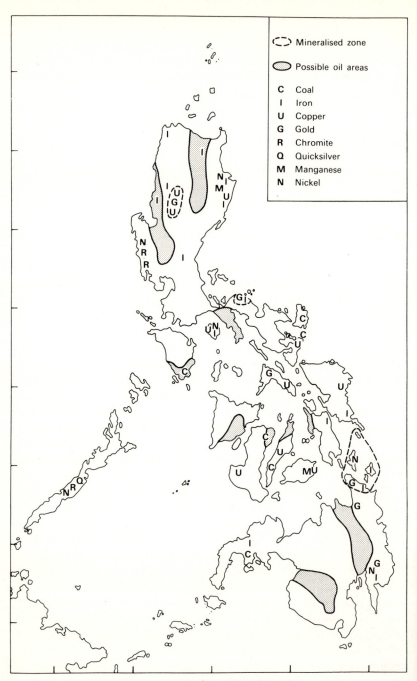

Fig. 19. Chief Mineral Regions

Tapian is a new mine inaugurated in late 1969; it serves a mill with a daily capacity of 15 000 metric tons and production is geared to the Japanese market. The Lepanto ore suffers from an excessive arsenic content whereas the value of the Toledo mine is enhanced by the presence of iron pyrites, which are converted to sulphuric acid at a nearby plant. A second mine at Toledo with a capacity of 28 000 metric tons of ore per day was inaugurated in 1970; it, like Tapian, is geared to the Japanese market and is designed to increase sales of copper concentrate by the Toledo complex to Japan by 60% (currently Japan takes three-quarters of Philippine copper exports).

A unique feature of Philippine mining is to be found in the province of Zambales, where one mine accounts for 80% of United States imports of refractory chromite. Annual output is in the order of 450 000 metric tons; proven reserves total some 10 million metric tons but many other deposits are known to exist in the vicinity. Metallurgical chromite also is mined in quantity in the Philippines.

An impressive range of other minerals are known to exist in the Philippines with interest currently focusing on nickel, not just from Surigao but also from deposits presently being surveyed in the provinces of Davao, Palawan and Zambales. Prospecting for minerals in the Philippines to date literally has scratched only the surface, for only one tenth of the country has been searched systematically. Furthermore, proven deposits generally are either too small in extent, or inferior in quality, or too inaccessible for profitable exploitation on a regular basis. Thus, for the moment at least, mining activities make but a small, though often locally significant, contribution to the rural scene.

References

1. Wernstedt and Spencer, *op. cit.,* pp. 183-4
2. J.E. Spencer, *Land and People in the Philippines: Geographic Problems in Rural Economy*, University of California Press, Berkeley, 1952, pp. 11-12.
3. Wernstedt and Spencer, *op.cit.,* p. 192.
4. For details of the numerous minor products see Wernstedt and Spencer, *op.cit.,* pp. 200-10.
5. Discussion of individual rural products has as its basis the detailed reviews contained in R.E. Huke, *Shadows on the Land: An Economic Geography of the Philippines*, Bookmark Inc., Manila, 1963. These have been updated by reference to the following key economic series:
   (a) *Fookien Times Yearbook*, Fookien Times, Manila.

- (b) *Philippine Economy Bulletin* (bi-monthly), National Economic Council, Manila.
- (c) *Philippine Weekly Economic Review*, Philippine Association, Manila.
- (d) *Quarterly Economic Review, The Philippines and Taiwan*, Economist Intelligence Unit, London.
- (e) Miscellaneous Philippine government (Bureau of Statistics) publications (generally data contained therein have been inadequate to make possible revision to Huke's maps which are based on latest available Census data).

6. For the most comprehensive study of shifting agriculture in the Philippines see: H.C. Conklin, *Hanunoo Agriculture in the Philippines*, F.A.O., Rome, 1957.
7. Wernstedt and Spencer, *op.cit.*, p. 291. To elaborate: 'The long-term contracts specify such items as the shares of sugar to be received by the growers and millers, procedures for estimating sugar content of delivered cane, stalkage that can be cut and delivered, and cane varieties to be grown; annual working agreements further specify cane-cutting patterns and schedules, delivery schedules for daily mill-quotes, and replanting patterns. The co-operative system involves integration of many complex elements between growers and millers not required in the same way in other countries.' (*Ibid*, p. 681.)
8. N.W. Schul, 'A Philippine Sugar Plantation: Land Tenure and Sugar Cane Production,' *Economic Geography*, Vol. 43, 1967, p. 168.
9. For details of the various types of corrals see Wernstedt and Spencer, *op.cit.*, pp. 234-5.

4
The Place

To the natural scientist the proximity of the Philippine Islands to the line of demarcation between Oriental and Australasian zoological regions (the so-called Wallace line) is perhaps the outstanding feature of the country's physical environment. To the geographer, however, the significance of the physical environment lies in the role played by its various elements in creating the 'place' in which the people conduct their life and livelihood. Consequently, this chapter is predominantly concerned with selected aspects of landforms, climate, soils and vegetation. It is the limits set by these elements, either individually or in conjunction, which man, governed by his present state of technological development and current economic conditions, is not disposed to challenge.[1]

Landforms

The Philippines is an archipelago whose seven thousand islands have emerged from the sea as a consequence of an extremely complex pattern of faulting, folding and volcanic activity. The fragmentation of the land mass means that coastal and offshore environments are of especial importance; hence, special consideration has to be given to, for example, the coral shoals of the Visayan Sea and the oceanic depths that parallel the Pacific littoral. Furthermore, an island's shape and size may play significant roles in determining the influence of marine elements: for example, Cebu, notwithstanding an absence of natural harbours, is nevertheless easily penetrated because of its narrowness; equally rugged Samar, despite possessing an indented coast, has been more difficult to penetrate on account of its compact shape and relatively large size.

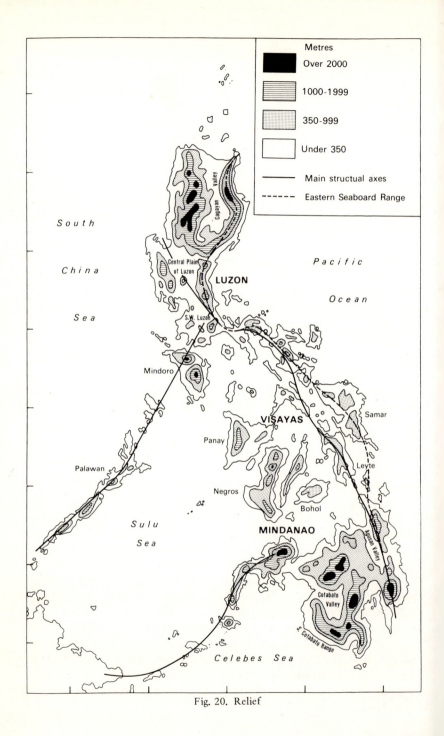

Fig. 20. Relief

The island environment results in considerable regional differences which can be highly localised as a result of elevation, aspect, etc., not to mention shape and size. However, it is rare for large contiguous natural regions to emerge. In this context, the absence of major drainage basins is of especial significance: only the Agusan (at 298 kilometres the longest river in the Philippines), the Cagayan and the Mindanao Rivers have created river valley environments of any size. Rather, a broad uniformity of place is apparent with frequent and often individualistic local variations from the norm being commonplace. As will be demonstrated below, such variations are basically a function of landform differences.

The geologic forces operative in the Philippines are common to all countries which make up the western seaboard of the Pacific Ocean: from Kamchatka in the north to New Zealand in the south, tectonic instability has created a distinctive pattern of landforms. Most geologists are of the opinion that the underlying cause of this instability is the imperceptible eastward drift of the Asian continent which is preceded by crumpling and uplift of its periphery, and in particular of depositional areas offshore. The Philippines is largely composed of these emergent marine deposits whose nature has subsequently been greatly altered by pressure and igneous influence. The rocks thus formed are generally Tertiary in origin (55%) but Mesozoic material (25-30%) may also be present especially where mountain cores have been exposed. Relatively recent vulcanism accounts for the bulk of the material outstanding.

The principal axis along which tectonic activity has occurred runs southeastward from Lingayen Gulf across the Bicol Peninsula and the island of Leyte before trending southward across eastern Mindanao (Fig. 20). Subsidiary axes include that running the length of Palawan Island on its way from Sabah to South-East Luzon and a parallel axis running through the Sulu Islands and the Zamboanga Peninsula. The line of these subsidiary axes can also be traced undersea, for comparatively shallow water marks their presence. Many minor axes have been produced as a result of the complexities of local mountain building created by the convergence of these three most prominent axes.

Lying generally to the east of the main mountain building

axis there is to be found a chain of deeply dissected mountain masses and associated uplands which extend the length of the island of Luzon and continue through Samar to Mindanao. This *Eastern Seaboard Range* represents the easternmost folding associated with the continental drift compression which has also produced offshore some of the world's deepest known ocean depths, notably the Mindanao Deep with a measured depth of 10 800 metres. From Cape Engano in the extreme north to the 'waist' of Luzon in central Quezon Province, the Eastern Seaboard Range is known as the Sierra Madre. Except to the west of Dingalen where a gap occurs with a maximum elevation of 245 metres, the mountains form a major barrier to east-west movement and are virtually unexplored. Peaks throughout range between 1100 and 1400 metres with maximum elevations in excess of 1750 metres. The central portion of the Range is in marked contrast to the Sierra Madre. In Southern Luzon and Samar it is replaced by relatively low-lying and fragmented uplands, the product of extensive faulting which has provided numerous exposures of ultra basic basement rocks of Mesozoic age. Rolling hills rather than rugged mountains are characteristic: in Samar the maximum elevation is in the vicinity of 850 metres, in the Bicol Peninsula elevations seldom exceed 500 metres. However, in the eastern part of the latter area a series of volcanic peaks, some with elevations in excess of 1500 metres, introduce a mountainous element to the landscape of which the near-perfect symmetrical cone of Mt. Mayon (2462 metres) is the culmination. Typical fold mountains again dominate the landscape in Eastern Mindanao. Here, the Eastern Seaboard Range (known locally as the Pacific Cordillera) can be crossed at elevations below 250 metres in the fifteen kilometres wide central saddle. On either side elevations in excess of 1500 metres are not uncommon especially to the south where one peak approaches the 3000 metre mark.

As the Eastern Seaboard Range usually rises abruptly from the sea the development of any coastal plain is restricted, being most evident in the central portion of the range where, however, fragmentation prevents the occurrence of any extensive areas. Isolated pockets of lowland separated by mountain spurs which reach the sea are characteristic; these are watered by small streams which plunge from the steep slopes and create tiny alluvial deltas at irregular intervals along the shore. In many places considerable coral growth is present, giving rise to both numerous reefs and white sand beaches both of which are constantly pounded by heavy surf.

Westward of the Eastern Seaboard Range no landform feature has other than local or regional significance. In Northern Luzon the western margin of the Sierra Madre consists of two major intermontane basins and one area of active vulcanism. The northernmost of the intermontane basins of Luzon is the *Cagayan Valley* which, although floored with 1000 metres of sediments derived from the adjacent uplands, possesses a rolling topography which is probably erosional in origin. The principal stream, the Cagayan River, and its major tributaries have laid down recent sediments but these floodplain areas are of no great extent compared with the total area of the Valley which is in excess of 5000 square kilometres. The other intermontane basin, the *Central Plain of Luzon*, is double the area of the Cagayan Valley. It, too, represents a major syncline occupied by the sea on several occasions. However, in contrast to the Cagayan Valley its surface is generally extremely flat and is largely covered with a layer of volcanic tuff derived from the volcanoes to the south.

The absence of local relief (the drainage divide between Lingayen Gulf and Manila Bay is only at an elevation of about fifty metres) hampers drainage and in certain areas, notably Pampanga Province, swampland is extensive. This is in addition to the extensive tidal swamps of Manila Bay and Lingayen Gulf which represent the present demarcation line between the basin's surface and the seas which periodically have inundated it. Both embayments, and Manila Bay in particular, are noted for calm waters and an absence of strong currents, a marked contrast to maritime conditions on the east coast.

The area of active vulcanism in *South-Western Luzon* is set in broad level areas of volcanic deposits of Tertiary and Quarternary origin and its numerous volcanoes indicate a major area of crustal instability set at the junction of the country's main mountain building axis and the subsidiary axis running from Palawan Island. The most infamous of the active volcanoes is Taal, situated on an island in Lake Taal which in turn occupies the crater of an older volcano estimated to have been more than 5000 metres high. To the north-east is the much larger Laguna de Bay, which is also volcanic in origin. In many respects the landform features of South-Western Luzon resemble those of eastern Bicol Peninsula discussed previously.

The mountains which form the western rim of the intermontane basins of Luzon exhibit marked dissimilarities of character. West of Cagayan Valley lies a complex mountain block created by both faulting and volcanic action. It is known as the *Cordillera Central* and has the Ilocos Range as its western outlier. Like the Eastern Seaboard Range its core consists of basic rocks of Mesozoic age but in this case extensively supplemented by Tertiary volcanic material. Maximum elevations are in the vicinity of 2750 metres and in total area the Cordillera Central constitutes the largest single compact upland mass in the Philippines. Between the Ilocos Range and the sea is to be found a coastal plain of varying dimensions being most pronounced to the north of Vigan where its width exceeds fifteen kilometres in places. It is also supplemented in Abra Province by several broad valleys of streams draining the Cordillera Central. These are almost encircled by the Ilocos Mountains. Flat land of only limited significance fringes the complex block of the *Zambales Mountains* and their southward extension, of Tertiary volcanic origin, into Bataan Province. Elevations rarely exceed 1250 metres but the uplands have been heavily dissected and consequently are very rugged. Despite a maximum width of only forty kilometres they provide an effective barrier to westward movement from the Central Plain of Luzon as does the Cordillera Central for the Cagayan Valley.

The Sibuyan Sea separates the Visayan Islands from Luzon and in turn it is separated from the much deeper and island-free South China Sea by the island of Mindoro which, with the adjacent island of Palawan, is associated with the subsidiary mountain building axis running from Borneo to South-Western Luzon. *Mindoro* itself consists of a broad anticlinal ridge of Mesozoic rock which forms the backbone of a rugged mountain divide whose peak elevations exceed 2500 metres. Rolling foothills surround the mountains and the coastal plains are also extensive. Like Mindoro, the backbone of *Palawan* is a rugged mountain range, though elevations are lower with peaks averaging between 1000 and 1500 metres. This long (415 kilometres) but narrow (average width 30 kilometres) island is characterised by extensive exposures of Tertiary limestone though in the central portion Mesozoic rocks are widespread. Unlike Mindoro, however, flat land is at a premium as the anticlinal ridge drops steeply beneath the sea to give depths in excess of 2000 metres within a few kilometres of the eastern and western shores.

A host of islands comprise the Visayas but, aside from the previously discussed island of Samar only five—Panay, Negros, Cebu, Leyte and Bohol—are worthy of individual treatment. *Panay*, which occupies the southern margin of the Sibuyan Sea, is composed of two anticlines separated by a synclinal basin which provides the only extensive area of lowland. Mesozoic rocks predominate in the main anticline which is to be found parallel to the west coast. This is an extremely rugged area forty kilometres wide where peaks average 1000 to 1250 metres in elevation and whose westward spurs break up the already narrow west coast plain. Eastward, the Central Plain of Panay extends completely across the island from the Sibuyan Sea in the north to Panay Gulf in the south. On the north-east it is bordered by the second anticline which is a much less prominent feature than its eastern counterpart, its maximum elevation being only 834 metres. *Negros*, which ranks immediately after Samar as the fourth largest of the Philippine islands, is dominated by four volcanic mountains only one of which is still active. Three, including the still active Mt. Canlaon (2438 metres), are to be found in the northern half of the island and are linked to the fourth, Magaso Volcano (1903 metres) in the south by a complex series of less prominent mountains running closely parallel to the east coast. Volcanic material has created the gently sloping, broad plains of North and North-Western Negros together with the area behind Dumaguete in the south-east. Level land is less extensive in the south-east which is occupied by a series of rugged hills extending to the coast in places.

Anticlinal structures are responsible for the creation of many of the Visayan Islands and no more so than *Cebu*, which is essentially a smaller version of Palawan both in shape and rock composition although it is composed of two parallel anticlines whose ridges possess average elevations in the order of 400 to 800 metres. The only significant area of lowland is in the vicinity of Cebu City. The backbone of the island of *Leyte* is formed of rugged mountains, chiefly of volcanic origin, whose peaks average 600 to 1000 metres in elevation except at its 'waist' where a low gap permits east-west movement. Unlike

Cebu, two important lowland areas exist: the larger is to be found in the north-east and is floored with Quarternary sediments many of which are of volcanic origin; the other lowland area lies in the vicinity of Ormoc City. The trio of islands which fringe the Camotes Sea is completed by the roughly circular island of *Bohol* whose origin can be traced to faulting and isostatic movement rather than folding. The main fault line parallels the south-east facing coast which is backed by the island's major ridge line whose highest elevations rarely exceed 750 metres. To the north-east of the ridge the gently sloping surface is composed entirely of Tertiary and Quarternary sediments, mostly limestone and coral, upon which subsequent erosion has created a typically humid area karst landscape in which poljes, sink holes, caves and underground drainage patterns are common. The dip slope surface continues beneath the surface of the Camotes Sea to produce a host of coral reefs, sandbars and mangrove swamps, features which are also to be found, though less well developed, in other parts of this comparatively shallow and sheltered body of water.

To conclude this review of the landform features of the Philippines it is necessary to return to the island of Mindanao whose extreme eastern portion forms the southern extension of the Eastern Seaboard Range. As in Luzon, a major intermontane basin lies adjacent to this range; in this case the *Agusan Valley*, measuring 175 kilometres from north to south and forty to fifty kilometres from east to west. The valley of the Tagum River which flows southwards to the Davao Gulf is separated from the Agusan Valley by a divide less than 200 metres in elevation. It, and the fifteen-kilometre-wide Davao coastal plain, can be regarded as an extension of the Agusan Valley, which is believed to occupy a broad syncline. The nature of this syncline's western border is obscured in part by volcanic activity which has created the north-south *Central Mindanao Highlands*. It is generally less well defined that its eastern counterpart but includes many impressive peaks, notably the active volcano of Mt. Apo (2953 metres) to the west of Davao which is the highest point in the Philippines. South of Davao elevations are less impressive with peaks averaging 1250 to 1750 metres compared with 1750 to 2500 metres in the northern and central portions of the Central Mindanao Highlands.

Terrain to the west of the Central Mindanao Highlands forms a complex pattern of plains, plateaux and volcanoes. Fringing the Mindanao Sea in the north, and indeed extending into it in the case of Camiguin Island, is an upland area dominated by a series of volcanic mountains many of which are still active. This upland area continues westward of Ozamis City to become more sharply defined as the *Zamboanga Cordillera*, an anticlinal structure associated with a subsidiary mountain building axis whose alignment can be traced toward Sabah through Basilan and the Sulu Islands (both of which were created by a mixture of volcanic and coral activity). Peak elevations on the Zamboanga Peninsula rarely exceed 1000 metres and small intermontane basins also are found. At the extreme eastern end of the Peninsula, however, volcanic action results in more massive peaks. Iligan and Cagayan Bays, with their generally restricted coast plains, are overlooked by a series of volcanic peaks, several of which exceed 2500 metres in elevation and appear to be even higher on account of their semi-isolated positions. At least a dozen of the thirty-odd volcanoes still are active, notably the series of six which comprise Camiguin Island in the extreme east.

Inland, the volcanoes give way to plateau-like terrain, part of which is occupied by Lake Lanao at an elevation of 750 metres. The *Central Mindanao (or Bukidnon) Plateau* is composed of basaltic lava flows interbedded with ash and volcanic tuff. The extensive plateau surface possesses an average elevation in excess of 600 metres but is interspersed with the various volcanoes which at times exceed 2000 metres in height. At several points in the north impressive waterfalls such as the Maria Christina Falls plunge over the plateau edge. Streams in the southern portions of the Central Mindanao Plateau follow a less spectacular course, draining to the Mindanao River which has created the *Cotabato Valley* with a length of 140 kilometres and a maximum width of 85 kilometres in its centre. Isostatic uplift in Pleistocene times has left its mark in numerous distributaries and swamps. A south-eastern extension of the Valley crosses a 300 metre watershed to reach the lowlands surrounding Sarangani Bay. A constriction of the Cotabato Valley occurs where the Mindanao River reaches the Moro Gulf in Ilana Bay. This is caused by the *South-West Coast Range* of volcanic mountains which sweep in a 180 kilometre arc from

Sarangani Bay in the south-east to Ilana Bay in the north-east. The mountains permit only a limited coastal plain in Southern Cotabato, a plain which is overlooked by peaks of varying height but which at times exceed 1500 metres.

Climate

By its location the Philippines comes under the influence of both the summer and winter airflows which cover South-East Asia. In spring and summer intensive heating of the surface of Mainland Asia creates a very strong and deep low pressure centre. This is sufficiently intense to attract the south-east trade winds across the Equator from the southern hemisphere thus creating, as a result of deflection by the earth's rotational movement, a drift of air from the south-west. In the vicinity of the Philippines this drift is somewhat modified, its strength and duration being less pronounced than at its centre over the Indian Ocean and its moisture content diminished through its merging with dry air attracted from northern Australia. During the winter months the pattern is largely reversed. The Asian land mass cools rapidly and an area of high pressure is created from the centre of which air flows out in a clockwise arc which in the western Pacific serves to augment the north-east trade winds. In consequence, their impact upon the east coast of the Philippines is noted for its force and regularity.

The key element in the above air mass pattern is the intertropical front which is the name given to the zone in which northern and southern hemisphere winds meet each other. Monsoonal influences tend to cause the intertropical front to become broken and intermittent, and over the Philippines it can best be regarded as a zone of relative low pressure causing local convergence of local winds. The typhoons which sweep north and west across the central and northern parts of the Philippines have their origin in the intertropical front, but such phenomena represent only one facet of the role played by the front in determining the most significant aspect of the Republic's climate—the rainfall pattern.

Rainfall is the most important climatic factor, the key to variations in vegetation, and hence agriculture. Thus, for example, less than 150 centimetres per annum normally requires the farmer to irrigate his rice or alternatively switch to growing maize. But, with warm moist air enveloping the archipelago throughout the year, dry weather, strictly speaking, is nonexistent. Average annual rainfall experienced by large parts of the Philippines is in excess of 200 centimetres and many areas receive over 250 centimetres; peak annual totals, in excess of 525 centimetres, are experienced at Baras on Cantanduanes Island.

Precipitation results from three causes: convectional activity associated with the intertropical front and the south-west monsoon it precedes, typhoons, and orographic activity. The intertropical front lies over the island of Mindanao for a greater period of time than elsewhere but its north-eastward movement at the time of the south-west monsoon and subsequent retreat causes its influence to be felt throughout the country. Rainfall from the typhoons in contrast is concentrated in those areas where convectional activity is restricted, that is, the north and east. Eastern parts, too, are most favoured in respect of orographic rainfall which is primarily derived from the strong north-east trades of the winter months.

The comparatively simple threefold pattern of rainfall distribution exhibits many local variations with rain shadow caused by relief features the outstanding factor (Fig. 21). Thus, all major areas of relatively low rainfall (under 200 centimetres), such as the Central Plain of Luzon, the east coast of Cebu and the Cagayan Valley, are protected from either or both of the monsoon winds. General Santos, the location with the country's lowest annual average rainfall (eighty-five centimetres) lies in a pocket almost entirely surrounded by mountainous areas. In such situations, the bulk of rainfall is derived from convectional sources.

The actual distribution of rainfall can be summarised as follows. Almost the entire east coast receives amounts in excess of 125 centimetres annually which is concentrated in the winter months when the north-east monsoon is at its peak, but is supplemented in the Eastern Visayas and Eastern Luzon from the passage of typhoons in the autumn (of the twenty typhoons which, on average, form in the West Pacific annually, three-quarters materially affect the weather of the Philippines and one third touch or cross the country.) Any lack of rainfall from typhoons in Eastern Mindanao is made up for by more pro-

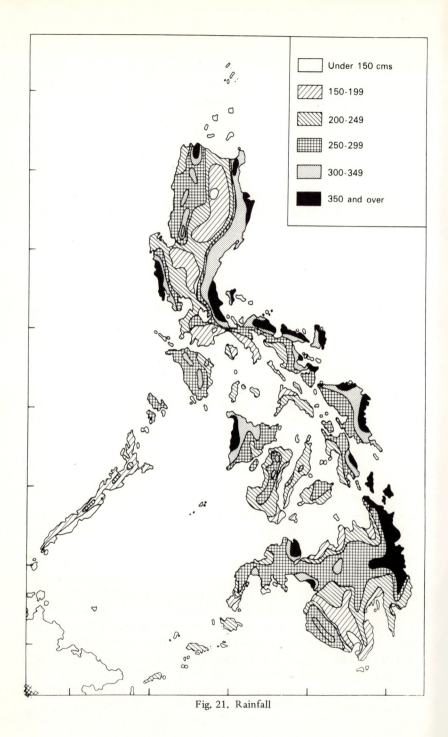

Fig. 21. Rainfall

longed activity associated with the intertropical front, which is largely responsible for the high rainfall experienced by most other parts of the island. The Central and Eastern Visayas are largely sheltered from the influence of the north-east monsoon and the typhoons. Their rainfall is received more in the summer months when the south-west monsoon is at its peak. Unfortunately many islands, and particularly Palawan and Cebu, are aligned parallel to the monsoon winds so their impact is reduced. A notable reversal of this situation occurs in Western Panay, and to a lesser extent in Western Mindoro. Here steeply sloping mountain areas lie athwart the south-west monsoon, producing ideal conditions for orographic cooling and subsequent very heavy summer rainfall. Similar conditions occur along parts of the west coast of the island of Luzon but this, apart from the eastern fringe, is primarily an area of low rainfall. Rain shadow effects and the limited presence of the intertropical front in late summer bring about this situation.

Despite the high rainfall densities experienced it must not be forgotten that they are associated with high temperatures, so that rainfall efficiency is drastically cut by evaporation. In view of this situation rainfall unreliability could pose problems. However, average departure from normal rainfall to be expected over the Philippines as a whole appears to be in the order of only 14%, being below this figure in Mindanao and above it in Southern and Central Luzon. Extreme variations of over 100% have been recorded but in all cases were surpluses caused by excessive typhoon activity.

As the Philippines lies wholly within the tropics, its latitudinal extent over more than 1500 km does not bring about any significant variations in *temperatures*. In the extreme north of Luzon, Laoag experiences a temperature range between the coldest and warmest months of only 3·8°C: in January the average temperature is 24·8°C and rises to a peak value of 28·6°C in May. Temperatures at Laoag fall back in the summer months as cloudiness increases with the onset of the south-west monsoon. Southward the temperature range steadily decreases and at the same time the maximum temperatures occur more commonly in the summer months rather than late spring as the increasingly early onset of the south-west monsoon reduces the opportunities for a coincidence of relatively cloudfree skies and a high elevation of the sun. Nevertheless, May remains the

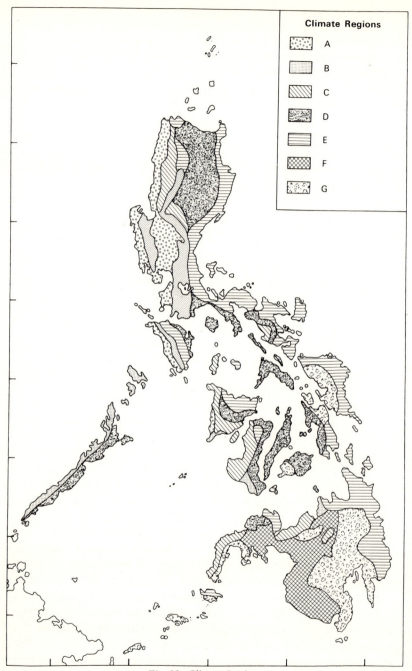

Fig. 22. Climate Regions

month for maximum temperatures as far south as the Visayas but August becomes the warmest month in Southern Mindanao. Here, an equatorial regime is experienced with very little variation in monthly temperatures: at Jolo for example, the coolest month is February with 26·0°C, only 1·9°C below the value for August which is the warmest month.

The temperature pattern described above provides no hindrance to plant growth at sea level, indeed recorded temperatures at this level have not been known to fall below 16°C, nor rise above 38°C except on very rare occasions. Increasing altitude provides opportunities for lower temperatures, but even at Baguio, at an elevation of 1500 metres, where the country's lowest temperatures are recorded, it is uncommon for values below 10°C to be experienced.

With all-year vegetation growth made possible by temperature conditions, the volume and seasonal distribution of rainfall plays the dominant role in determining the regional variations in climatic conditions throughout the Philippines (Fig. 22). In turn, the pattern of agriculture responds to such regional variations: for example, production of the staple crop—rice—can

TABLE 15
Classification of the Philippine Climate

Traditional	Revised
1. Dry winter and spring, wet summer and fall.	A. Long low sun dry season; 5 or 6 months with less than 5-10 cm of rainfall per month.
	B. Intermediate low sun dry season; 4 months with less than 5-10 cm of rainfall per month.
2. Short dry season of 1-3 months.	C. Short low sun dry season; 1-3 months with less than 5-10 cm of rainfall per month.
	D. Short high sun dry season; 1-3 months with less than 5-10 cm of rainfall per month.
3. No dry season, pronounced winter rainfall.	E. All months with 5-10 cm or more rainfall; wettest low sun month with at least 3 times the rainfall of dryest high sun month.
4. None.	F. All months with 5-10 cm or more rainfall; wettest high sun month with at least 3 times the rainfall of dryest low sun month.
5. No dry season and no pronounced maximum rainy season.	G. All months with 5-10 cm or more rainfall; wettest month has less than 3 times the rainfall of the dryest month.

Source: R. E. Huke, *Shadows on the Land: An Economic Geography of the Philippines,* Bookmark Inc., Manila, 1963.

range from double cropping without reliance on irrigation to single cropping dependent on irrigation, differences in approach which are of major significance to the local population.

The following classification is based upon that developed by Huke[2] which is designed to reveal variations in the yearly pattern of rainfall, the period of maximum rainfall and the length and season of any dry spells (Table 15). In the author's opinion it is the classification which best expresses the nuances of the Philippine climate.

Climate type A is marked by a long dry season and a very pronounced wet season. In other words this is a normal monsoon climate found in many parts of South-East Asia. In the Philippines the period of heavy rainfall occurs between April and September usually resulting in seven wet months (over six centimetres each month) and five dry months. However, in parts of Luzon there may be six dry months resulting in serious water shortages in the months of March, April and May which coincide with the period of highest temperatures. About 12% of the country experiences this type of climate and it includes such important areas as the Ilocos coast, much of the Central Plain of Luzon, the coastal lowlands of Antique, and areas to the south of Manila Bay.

Climate type B is characterised by a shorter and less severe dry season, lasting only four months. The extended rainy season results in a reduction of the temperature range, fewer water shortages and more luxuriant vegetation. Like type A it also covers about 12% of the Philippines. In Luzon it is characteristic of those parts of the Central Plain and its environs not experiencing type A, that is, a large area to the south-east of Manila Bay and the Zambales Mountains. Other areas include the western portions of the islands of Palawan and Mindoro, and pronounced rain shadow areas in Mindanao such as Zamboanga City.

Climate types C and D both are marked by a short dry season of one to three months. However, in the case of type C the dry season occurs between October and March whereas the dry season for type D occurs in the other half of the year. The reduction in the length of the dry season now permits a climax vegetation dominated by dipterocarps thus displacing the molave forest characteristic of types A and B. Type C climatic conditions are experienced by about 10% of the Philippines, in areas scattered throughout the length, though not the breadth, of the country. Type C climatic conditions are absent from the eastern half of the country for the north-east trade winds do not permit any dry season between October and March. Elsewhere, type C conditions are experienced in areas such as the western portions of Panay and Negros which are sheltered from the north-east trade winds and which are exposed to the influence of the south-west monsoon. Even in Luzon where the influence of the monsoon is lessened, localities such as the south-facing slopes of Mountain Province, which are positioned to secure copious rainfall at other times of the year, also experience this type of climate. The type D climate with its April-September dry season, affects almost one fifth of the Philippines. The whole of the Cagayan Valley experiences this climatic condition as do such localities as the islands of Masbate, Cebu and most of Bohol. These represent areas which benefit relatively little from the south-west monsoon, receiving instead most of their rain from typhoon activity and the north-east trade winds.

A marked seasonal period of heavy precipitation coupled with an absence of any dry season is typical of both climatic types E and F; the former is the most widespread of climatic types under discussion, covering 22% of the area of the country. Here, the period of heavy rainfall can be attributed to the north-east trade winds and, in most cases, typhoon activity. The entire eastern seaboard of the Philippines experiences such climatic conditions as do north-east-facing slopes in Panay and Mindoro. Heavy dipterocarp vegetation flourishes under these conditions. In contrast to the large extent of type E, type F occupies the smallest area of the Philippines' different climates, less than one tenth in fact. Being characterised by exceptionally heavy rainfall between April and September, it is to be found in that part of the Philippines most affected by the south-west monsoon, i.e., Mindanao, and in particular the southern and western sections of the island. The final climate, type G, is characterised by heavy rainfall throughout the year but with no marked seasonality. As it is associated with small temperature ranges it is the most monotonous of the Philippine climates. It covers about 15% of the country and is to be found in two areas—northern Leyte and western Samar, and South-Eastern Mindanao.

Soils

The youthful geologic nature of the Philippines causes soils to be strongly influenced by their parent materials. Andesites, basalts and agglomerates are the most widespread of the parent materials (Table 16) but alluvium, which forms the basis for the most productive soils, covers a respectable 15% of the land surface. The distribution of Philippine soils (Fig. 23), however, highlights the fact that in many parts of the country such as Palawan there can be found a predominance of skeletal soils. In fact, throughout the Philippines such inferior soils occupy one fifth of the total area. Nevertheless, in comparison with other tropical areas Philippine soils overall rank among the more fertile, due in no small measure to the extensive occurrence of coralline limestone and volcanic materials. Unfortunately their inherent fertility, moderate as it is, has been seriously affected by inefficient agricultural practices leading to serious soil erosion.

Alluvium supports both the most intensive agriculture and the highest Philippines rural population densities. Fine textured alluvial soils, in particular the San Manuel silt loam, are most favoured in view of their suitability for rice culture. Most notable concentrations of such soils are to be found in the Central Plain of Luzon, the Cagayan Valley and the Cotabato Valley (Mindanao). Alluvial soils of coarser texture are more permeable but are subject to rapid leaching and consequently are of relatively low fertility. Nevertheless, despite their unsuitability for lowland rice, such soils are of considerable agricultural value being favoured for the production of, for example, sugar, abacá, coconut, and tobacco. The Central Plain of Luzon, and especially the Pampanga and Tarlac areas, possesses large areas of this type of soil as do Negros Occidental and the Koronadal Valley (Cotabato). Unfortunately, the as yet unsolved problem of alkalinity combined with low fertility reduces the value of the Koronadal Valley alluvium.

Both *shale* and *sandstone* occur sporadically throughout the Philippines with shale the more common as a soil parent material (under-lying about one tenth of the total land area or double that for sandstone). Variations in rock hardness, especially in the case of shale, gives rise to a variety of soil series which usually possess a clayey texture. Such soils are difficult

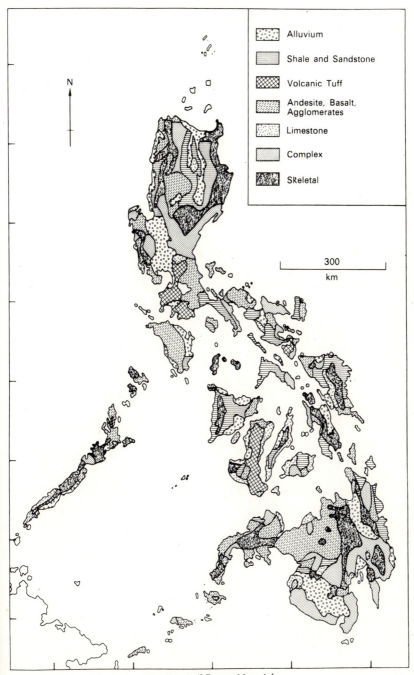

Fig. 23. Soil Parent Materials

TABLE 16

Principal Features of Philippine Soils

Parent Material	Areal Extent (%)	Colour	Character	Areas of Concentration	Land Use
1. Alluvium	15	Various, depending on age and source of parent material.	*Recent deposits*: coarse gravel and sand, silts and clays. Low fertility, difficult to work. Soils are noted for their rapid permeability and good drainage. San Manuel, Zuingua and Umingan series are exceptions—more fertile, less permeable and can support rice. Consist of light brown loose, friable loams.	Central Plain of Luzon, Cagayan Valley, Rio Grande Basin.	Sugar, coconuts, fruit, vegetables, rice (if fertility high and permeability low).
			Fan and terrace deposits: fine textured, clayey and sandy loams. Relatively fertile. Internal drainage poor and permeability very low. All soil types above respond well to fertilisers and green manures.	As above.	As above but sugar only if well drained.
2. Shale	10	Dark red-reddish brown.	*Non-calcareous* (Palompan, Maasin and Alimondian series): low fertility.	North and South-East Luzon, Leyte and Panay.	Grass, forest.
			Calcareous (Sevilla and Lugo series): fertile. All soil types above are fine textured with low permeability—solum 15-60 cm.	Cebu, Bohol, Leyte, Negros Oriental.	Fruit, maize, root crops.

3. Sandstone	5	Brownish grey-grey.	Coarse granular sands with clayey subsoil, the product of leaching. Little humus—infertile.	Bohol.	Forest and grass.
4. Volcanic tuffs	2	Brown-black.	Depth considerable (over 3 metres), top horizon, 30-50 cm low fertility, deficient in phosphorus. Loose coarse granular structure unless water laid, then fine texture, shallow, low permeability.	South-West Luzon.	Sugar, rice, coconuts, fruit vegetables.
5. Andesites, basalts and agglomerates	21	Red-reddish brown.	Lateritic, friable, acidic. Porous. Depth similar to volcanic tuffs. Low in calcium and phosphorus. Typical series—Antipolo, Guimbalaon and Luisiana.	Luzon, Mindanao, Western Visayas, Mindoro.	Abacá, coconut, grass, forest.
6. Limestone	13	Black.	Faraon series typical—rendzina type developed on soft coraline material. Shallow, friable, very porous, high organic content and concentration of alumina and iron.	Cebu, Bohol, Samar, Negros, Northern Mindanao, Cotabato.	Maize, coconuts.
		Red-reddish brown.	Bolinao series typical—terra rosa type developed on harder parent material. Profile 10-12.5 cm deep with high residual clay content. Interspersed with limestone, gravel and stone.	As above.	Maize, coconuts, bananas, citrus, forest.
7. Others	34[a]	Various but typically grey-greyish brown.	Metamorphic parent materials predominant, giving rise to shallow soils with low fertility. Deficient in organic matter, phosphorus and potassium.	Eastern Seaboard Range, Palawan, Negros, Leyte, Central Mindanao.	Forest and grass.

[a] Includes unsurveyed areas.

to manage; when wet they are plastic and sticky but become hard and crack when they dry out. The resulting low yields are accentuated by the inherent infertility of the soil. An exception to the above occurs in parts of Bohol and Cebu where the calcareous nature of the shale gives rise to a fairly rich and productive soil. Here too, however, the sloping terrain with which it is associated and its low permeability creates a severe soil erosion danger.

Volcanic tuff is very limited in extent as it is the product of geologically recent volcanic eruptions. Most is to be found in the provinces surrounding Manila and in Albay. Soils which develop from this material are shallow in depth, fine-textured and very plastic in consistency. These soils have been extensively worked with the result that soil erosion has become a serious problem. This is especially so in Batangas where some farmers are cultivating on the actual bedrock and forming 'soil' by pounding the adobe type rock to a suitable fineness.

Andesites, basalts and agglomerates form a group of parent materials, derived from igneous rocks of volcanic origin apart from those which are the product of recent eruptions, which are responsible for a wide variety of soil types. They are the most common of Philippine soils with especially large concentrations occurring in Mindoro, Luzon (Quezon and Mountain Provinces) and Mindanao (Lanao and Misamis). Where soils remain *in situ* they tend to be shallow and generally poor in fertility especially where the parent material possesses a high degree of hardness or has suffered metamorphosis. Where material from igneous rocks is washed to lower areas the sediments are fairly deep and the columnar structure of the resultant lathosols enables them to possess a high degree of permeability despite their clayey texture. To counteract the inherent poor quality of these soils heavy applications of lime are necessary not only to supply calcium to the crops but also to correct soil acidity.

Soils developed from *limestone* are to be found throughout the length and breadth of the Philippines, representing 13% of its land area. They are most prevalent in the Visayas. Most parent material is coralline limestone and the resultant soils are characterised by a marked degree of friability despite their granular clayey texture. Rapid soil development leads to a general absence of the B horizon, a feature accentuated by the general shallowness of the soil. Soils are highly susceptible to soil erosion and 'bedrock farming' occurs in Cebu where limestone soils achieve their greatest concentration.

The influence of the youthful geologic character of the Philippines has meant that soils surveyors are unable to categorise the soils of approximately one third of the total area of the country. This situation arises from two principal causes: the rugged nature of the terrain, which over large areas only permits the establishment of shallow immature skeletal soils, and the fragmentation of surface geology as a consequence of recent tectonic activity, which makes for a variety of parent materials and hence a complexity of soil types.

Vegetation

Compared with other countries, climate variations are not marked in the Philippines nor is any one type of soil dominant over large areas. This absence of any decisive regional influence leads to an overall lack of complexity in the natural vegetation. At the same time the importance of relatively minor differences in climatic and soil conditions determining the detailed structure of the vegetation cover is accentuated. Thus, for example, mangrove development is at its maximum where swamp soils possess a high degree of alkalinity, and molave displaces dipterocarp forest on shallow and rocky terrain. Elevation, too, produces important modifications (Fig. 24).

As in many other parts of the world the natural vegetation of the Philippines has been greatly modified by man's action. Over one third of the area of the Philippines is now cultivated or utilised for urban and other purposes (Table 3), leaving only some 20 million hectares outstanding. Within this area the forestlands are more than five times more extensive than the grasslands. Shifting cultivation, or *kaingin* as it is known locally, also has had a profound effect upon the nature of the existing vegetation cover.

Despite the influence of man, the forestlands are widespread (see Fig. 10) and include substantial areas of largely untouched primary forest. Locations most suited to the growth of primary forest are to be found in well drained valleys below 750 metres on the lower slopes of mountains. Here, all three storeys are present and the ground covering is absent or limited to a few

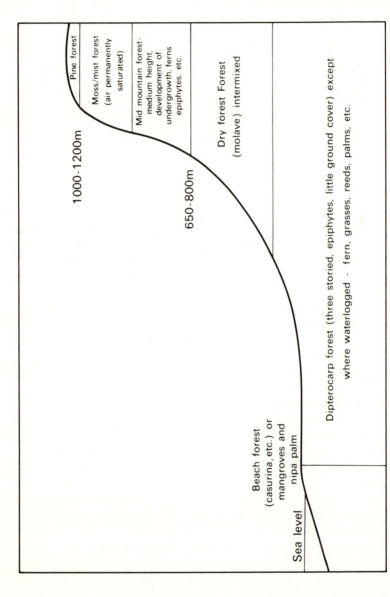

Fig. 24. Outline Profile of Climax Vegetation resulting from changes in elevation

terrestrial herbaceous species. In regions where the dry season is not pronounced, epiphytes, vines and climbing palms are usually abundant. Members of the evergreen dipterocarp family form the principal species of tree in the upper storey. They include the *lauan*, or Philippine mahogany, together with *apitong, guijo,* and *yakal,* a trio of equally large trees which yield much harder wood. However, hardwoods *par excellence* are to be found in the molave forests which develop where rainfall is less prolific and markedly seasonal and where soils are shallow and infertile. The molave forest can be classified as semi-deciduous and is dominated by members of the legume family, the most well known species being *molave, ipil* and *narra.*

With increasing altitude, the bigger trees thin out but epiphytes and terrestrial species increase in numbers. In high and very mountainous regions the combination of heavy rainfall, steep slopes, shallow soil and high winds causes trees to be mostly short bowled and covered with mosses, liverworts, ferns and epiphytic orchids. In the mountainous regions of Northern Luzon and Mindoro this pattern of vegetation gives way to pine forest. *Benguet* and *tapulau* pine are the most common species and often exist in practically pure stands, though open and scattered in extent. Pure hardwoods may be intermixed in areas protected from fire.

Generally, the distribution of the forestlands discussed above is patchy, being concentrated not so much in those areas most conducive to growth but in rugged areas distant from centres of population. The relatively undeveloped islands of Mindanao, Mindoro and Palawan fall into this category as do the more inaccessible parts of Luzon such as Northern Mountain Province and the Sierra Madre. Two other types of largely untouched primary forest exist in the Philippines though their areal extent is limited. Mangrove forest occurs along the tidal flats at the mouth of streams and along the shores of protected bays. Various species of Rhizophoraceae are dominant but *nipa* palm is also widespread in some areas. On sandy beaches above the high tide limits may often be found a mixture of many varieties of small and medium height semi-deciduous trees growing behind a frontal zone in which pandanus plays a conspicuous role.

In forestlands where the influence of the *kaingineros* has

been felt large areas of second growth forest occur. In time, natural regeneration will cause the cover to approach the virgin forest it replaces. Initially, however, the effect of *kaingin* is to destroy the forest canopy and encourage the growth of many terrestrial species, lianas and epiphytes which surround the surviving upper storey trees and inhibit new arboreal growth. The increased susceptibility of the area to forest fires introduces an additional inhibiting factor.

Over 3 million hectares of brush, *cogon*, or open grassland exist in the Philippines, the great majority of which has been created by man and particularly by the use of the *kaingin* system. The latter, which involves the felling and burning of the original vegetation cover followed by a few years' cultivation and subsequent abandonment, has enabled man to destroy the original forest cover. Where population pressure forces the local inhabitants to return to previously abandoned *kaingin* areas, natural regeneration of second growth forest is checked and, if the area is continually being brought back into cultivation, a grassland regime is established.

Cogon and associated varieties form the most dominant plants of the grasslands. As Spencer[3] points out:

'These tropical grasses form a very heavy turf, spread through root growth as well as seed dispersal, and are not killed by the annual fires. Although they do prevent heavy erosion, the periodic burning and the very nature of the grasses themselves are not helpful in rebuilding soils. Hence many of the grass-covered uplands today possess soils that are seriously impoverished and which nature is not aiding man to rebuild.'

Unfortunately the grasslands have little pasture value unless further burning is employed to stimulate new growth. Despite repeated burnings, quick growing trees exist, notably the *alibangbang* and *binayuyu*. Such trees have well developed roots and sprout readily from the base of the stem after the upper portion has been killed. A parkland aspect is thus characteristic of the grasslands.

The grasslands of the Philippines possess no marked areas of concentration, apart from the Bukidnon-Lanao plateau and Southern Cotabato. In general they occupy the rolling uplands which form an area of transition between the intensively cultivated lowlands and the rugged inaccessible mountains. The island of Masbate is largely covered with *cogon* grassland and extensive areas can also be found in, for example, Mountain Province, Samar, Western Mindoro and Panay.

Not only have man's activities been largely responsible for the forestland-grassland dichotomy but also he has greatly accelerated soil erosion—a physical force which, even when operating under natural conditions, can be very potent in a tropical land of marked relief such as the Philippines. Such has been the impetus given by *kaingin* and associated methods that some three-quarters of the country's farmland is affected by soil erosion to a greater or lesser degree. It is no coincidence that those islands least affected—Palawan, Mindanao and Samar (Fig. 25)—are characterised by recent settlement and low population densities. Soil erosion reaches its peak in intermediate piedmont zones:

'These gently sloping lands, with slopes ranging between 3% and 8% and occasionally reaching up to 15%, have been occupied mainly by marginal or ephemeral farmers. Particularly in the densely populated areas where fertile lowlands are both limited in area and already intensively utilised, the population has faced the problem of expansion into the adjacent piedmont and foothill lands. Unfamiliar with contour plowing and planting, or unwilling to put these methods into practice, the farmers have usually planted these sloping lands to row crops. In areas where the soil is extremely shallow or contains a hardpan layer, thereby impeding percolation of rainwater, a very rapid run-off often ensues. To cultivate these latter soils properly, deep plowing must be done so that the impervious layer is broken and free movement of soil water restored, an operation that is difficult to accomplish with only the power of a single carabao.'[4]

The famous rice terraces at Banuae, Mountain Province, might suggest that Filipinos are skilled at soil conservation. Unfortunately, the reverse is true, exemplified by conditions in Cebu where, as early as the late Neolithic, perhaps half the climax vegetation had been removed.[5] Climatic and soil conditions have made a significant contribution to the degree of soil erosion in Cebu, but the island is still worthy of study for it clearly indicates both past trends and future consequences of ill-advised farming methods in the Philippines:

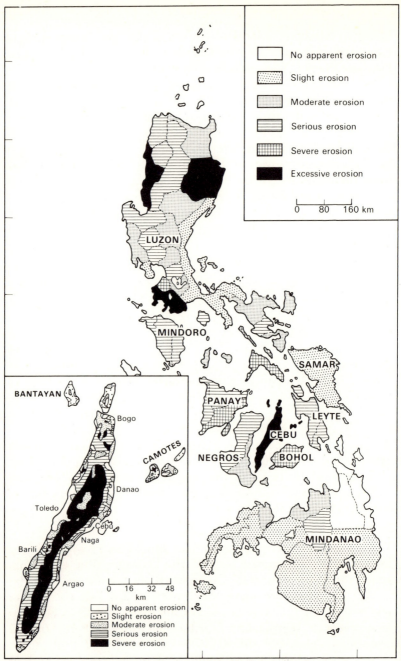

Fig. 25. Soil Erosion, by Provinces (Cebu Province in detail as inset)

'On this island fully 90% of the total soil cover now suffers from various degrees of erosion, and nearly 60% of the land is classified as severely eroded. On more than one third of the island's land all of the topsoil has been removed. On Cebu a large proportion of the land under cultivation lies on very steep slopes; many slopes range in steepness up to forty-five degrees. The increasingly crowded population of the island has forced the continual occupance of many of the steeply sloping lands in spite of the many obvious hazards. In order to bring these lands under cultivation almost the whole of the natural forest cover has been removed. To further compound the soil erosion problem, corn grown as a row crop is continuously planted on the steeply sloping fields. During the preparation of fields, the Cebuano farmer usually plows up and down the slope, thus creating natural furrows between the rows of corn which are, in essence, miniature gullies or rivulets. The downpours so typical of the tropical climates simply enlarge upon the drainage system thus provided. The net effects of severe erosion on Cebu include removal of the fertile topsoil from much of the land, elimination of considerable farmland from production because of forced abandonment of the severely eroded fields, and the frequent destruction of fields and crops on the lowlands below. Removal of the more water-retentive topsoil and the natural vegetation cover has led to an increased susceptibility to edaphic drought in the badly eroded areas of Cebu, and it is very apparent that many plants now suffer from even a short dry period.'[6]

References

1. This and succeeding sections are based initially upon R.E. Huke, *op. cit.,* pp. 1-71.
2. *Ibid*, pp. 44-51.
3. J.E. Spencer, *Land and People in the Philippines: Geographic Problems in Rural Economy*, University of California Press, Berkeley, 1952, p. 31.
4. Wernstedt and Spencer, *op.cit.*, p. 81.
5. C. Vandermeer, 'Population Patterns on the Island of Cebu, The Philippines: 1500 to 1900', *Annals of the Association of American Geographers*, Vol. 57, No. 2, June 1967, p. 320.
6. Wernstedt and Spencer, *op.cit.*, pp. 78-81.

5
The People

The United States Supreme Court once ruled that a Filipino is not of Caucasoid origin, and on the other hand, the state of California has ruled that he is not of Mongoloid origin.[1] To the geographer, this pre-occupation with racial characteristics is of limited relevance to his study of a particular group of people. Instead, attention is focused upon those aspects which make a noticeable contribution to the manner in which that group of people occupy the area of the earth's surface in question. In the case of the Philippines, an outline of racial characteristics is but a prelude to a review of more fundamental or dynamic aspects of the Filipino people.[2] Thus, for example, a review of demographic features reveals the pattern of occupation, while consideration of social attitudes points up the way in which this occupation has been organised. It is these and other related human influences which, either individually or in conjunction, have by their interaction with physical features created the geography of the Philippines.

It is also worth pointing out that there are additional, often intangible, attributes of the people of the Philippines that should be taken into account, particularly where regional evaluations are attempted. To quote Spencer:[3]

'There is the expansive generosity and love of good living of the inhabitants of Iloilo Province of southern Panay as contrasted with the canny thrift of the natives of Bohol Island a bit further east in the Visayas. The Ilocano from the north-west-coast provinces of Luzon willingly goes as a poor but thrifty colonist to many parts of the islands to obtain a better standard of living than he had in the overcrowded Ilocano territory. But the Tagalog from near and south of Manila willingly goes outside his home region only as a government official or merchant's representative.'

Certain characteristics of the people of the Philippines merit special attention by the geographer. Their very numbers is an obvious starting point, a state of affairs which applies to most Asian nations but especially to the Philippines whose population growth rate leads all others. A close second in importance is the land tenure system, for in an agricultural nation such as the Philippines not only does the livelihood of most inhabitants depend on the land but their way of life often is determined by the nature of their title to it. Finally, social attitudes need to be singled out for attention; Filipino behaviour patterns possess various unique features and these, like population pressure and land tenure, have a considerable impact on patterns of life and livelihood in the archipelago. First, however, it is desirable to review briefly the other fundamental features of the people who make up the nation of the Philippines.

Basic Features

Race. The dominant racial stock in the Philippines is Malay in nature. Preceded by negritos and other aboriginal races, successive waves of migration from Indonesia and mainland South-East Asia laid the foundation of the present population. More recently, Chinese, Spanish and American strains, in that order, have been introduced. Once established, Malay dominance of the islands became permanent; racial mixing may have produced a definite Filipino blend but to the uninitiated the Filipino closely resembles his neighbours in Indonesia and Malaysia.

As a result of the dominance of Malay stock, persons of other races make up only a small proportion of the total population. By far the most important are the Chinese whose numbers are in excess of 100 000 persons. Chinese traders are believed to have visited the Philippines about 2000 B.C. and it is only in the recent years that attempts have been made to curtail their ever-growing influx. The close knit community life of the Chinese is largely responsible for their success in retaining their racial identity though this has not prevented them making a notable contribution to the creation of the Filipino blend also.

Of the other racial groups in the Philippines, the negritos now are confined largely to the more remote uplands of the Visayas while, of the other aboriginal groups, only the Ifugao of Northern Luzon and the sea gypsies of South-Western Mindanao survive in any appreciable numbers (the discovery, in 1970, of an hitherto unsuspected stone age tribe—the Tassdays—living in a remote part of the rainforest of Cotabato represents an extreme example of the tenuous hold most minority racial groups now have on available living space). Few Spanish have retained their racial identity, a not unnatural state of affairs in the face of up to 400 years of residence and the fact that it was the exception rather than the rule for the Spanish colonists to be accompanied by their womenfolk. A small Indian community exists in Manila but it numbers barely 2000 persons; like that of the Chinese it is essentially inward-looking and close knit. In the last century, the displacement of Spanish influence by the Americans led to a sharp increase in the western expatriate population and this was further swollen as a result of military activity associated with World War II. This element of the population rarely has been permanent in nature but its size, in excess of 20 000 persons at present, makes an important contribution to the total non-Malay population.

Not many Filipinos can claim to be one hundred per cent Malay, particularly amongst the better off sections of society. All of the non-Malay peoples have contributed to a greater or lesser degree to the racial mixing which has created the Filipino blend. Original intermixing with the aborigines was but a preliminary to extensive *mestization*. Chinese *mestizos* are the most numerous, accounting for about one tenth of the Filipino population, but a Spanish strain is also prominent. Both have resulted in subtle but nonetheless definite modifications to the original Malay racial characteristics. More recent *mestization*, principally the result of an American admixture, has affected significant numbers of the present population but it is doubtful if in the long term it will modify the Filipino blend to a noticeable degree.

Language. Geographical fragmentation, caused by the dual barriers of the sea and the mountains, has led to a high degree of linguistic diversity in the Philippines (Fig. 26). Population mobility and a high degree of literacy (in excess of 75%) are increasingly blurring linguistic boundaries and minimising the

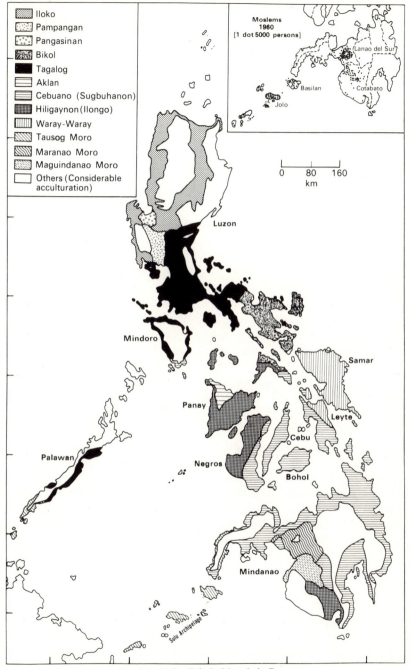

Fig. 26. Major Ethnic-Linguistic Groups

significance of the less common dialects. Nevertheless, language remains one of the more significant regional variants of the nation's social structure.

Eight major dialects are spoken in the Philippines and they are supported by seventy or so lesser native linguistic groups which are the most prominent in the highlands and similar remote areas. Five of the major dialects originated in Luzon and nearby islands and the remaining three in the Visayas. Internal migration has caused considerable diffusion of the basic language pattern. In Manila, Tagalog speaking persons have to contend with growing minorities speaking all the major dialects of the country, while in Mindanao the indigenous dialects face increasing competition from, for example, Cebuano, Ilocano and Tagalog, as the immigrant population grows in size. Tagalog, by virtue of being associated with the nation's capital and outstanding centre of population, is the dialect most widely understood (by 44.4% of the population in 1960). Before long Tagalog will be understood by the majority of Filipinos for Tagalog speaking people are in general better educated, have a better literature, possess a somewhat richer language, and wield considerable political strength. Thus, Tagalog was designated the National Language in 1937. The present National Language (Pilipino) has been somewhat modified but remains basically Tagalog. Pilipino is a required course at all levels of instruction so that, as a result, the younger generation is becoming steadily more proficient in Tagalog.

Non-indigenous languages spoken in the Philippines range from Chinese, used only within the Chinese community, to English which has become the *lingua franca* of the islands, displacing Spanish. English is not only the language of commerce and of the professions, a required language of instruction (with Pilipino and the local dialect) and the language of the principal newspapers, but also it makes possible conversation, albeit at times in a garbled form, with almost any Filipino in almost any place, however remote, in the country (in 1960, 39.5% of the population were classified as able to speak English proficiently). Unlike English, Spanish even in its heyday was confined to the upper echelons of society. Today it is still a language of consequence in the Philippines, although ability to speak it well has nowadays become a social rather than a practical asset.

Religion. Four centuries of Spanish colonisation, inspired as much by religious as economic and political motives, have left perhaps their deepest imprint upon the religious beliefs of the nation. Over 80% of the population are Christians and 90% of Christian Filipinos are Roman Catholics. Of the non-Christians most are Moslems or pagans, the former being located primarily in Southern and Western Mindanao and the latter in the remoter uplands of the larger islands. Most Christians who are not Roman Catholics belong either to the Philippine Independent Churches (*Aglipayan* and *Iglesia ni Kristo*), which are indigenous variants of Roman Catholicism, or to a wide range of American based Protestant groups such as the Episcopalians, Baptists and Methodists (Table 17).

TABLE 17
Principal Religions of the Philippines, 1960

	Membership ('000 persons)	%
Roman Catholic	22 686	83.7
Philippine Independent Churches[a]	1685	6.2
Protestant Groups	785	2.9
Moslem	1317	5.0
Others	604	2.2
Total	27 077	100

Source: Bureau of Census and Statistics.

[a] Mostly members of the Aglipayan Church (1 414 000 persons).

Religious fervour is still a very potent factor in the social and cultural life of the Philippines. Extremist actions are not uncommon: Moslems running amuck and bent on slaughtering all Christians in sight, and Catholics being temporarily crucified occasionally hit the headlines. But more fundamental is the day-to-day influence of religion. Regional variations are limited in number but have considerable local impact while throughout the way of life is conditioned by religious beliefs.

Roman Catholicism, despite the pressures of modern civilisation, still exerts a dominant influence upon the lives of many Filipinos. Unfortunately, as in certain other underdeveloped parts of the world, the role of the Church suffers from an over-emphasis upon the spiritual shortcomings of the community to the detriment of its material shortcomings. The Church, for example, has played but a minor role in agitation for land reform, doubtless in part as a result of its considerable financial dependence upon the principal landowners. In matters of national policy the voice of the Church often either is muted or is ultra conservative but at the local level a very different state of affairs exists. The parish priest may be a more powerful, certainly a more respected, figure in the community than the local politician, for his role in it extends beyond officiation at baptisms, confirmations, marriages and funerals, important as these events are in the social life of the average Filipino. Social events, particularly in rural areas, are rarely a success without the blessing of the parish priest who may also have had a major role in their organisation. Some such events, of which the *fiesta* is the prime example, will in any case be religious celebrations. The priest may participate in public life also by acting as arbiter in many personal and communal matters, especially those relating to the provisions of the Civil Code in matters touching marriage and divorce which coincide almost completely with corresponding legislation in Catholic Church Law.

The all-pervading influence of the Roman Catholic Church is exemplified by the state of affairs in education and politics. Religious intervention in both spheres is contrary to the provisions of the Philippine Constitution which provides for the separation of church and state. Nevertheless, religious instruction is commonplace in schools while intervention in politics may range from the parish priest indirectly canvassing support for a local councillor to full page political announcements in Manila newspapers, paid for by the Catholic archbishops.

In certain fields, the influence of the other religious groups can be strong also. The Philippine Independent Churches, with their combination of catholicism and nationalism (national heroes, for example, were added to their calendar of saints), for a time appeared likely to cause major defections from orthodox catholicism in Northern Luzon. The Protestants, however, can claim that finance rather than nationalism would appear to achieve more lasting success. Thus, by sponsoring educational institutions or hospitals, such as the Mary Johnston Hospital in Manila (Methodist) and the Central Philippine University in Iloilo (Baptist), the Protestant groups have been able not only to markedly improve existing facilities but also to secure a sizeable number of converts. Recent construction by *Iglesia ni Kristo* of a series of impressive and well designed churches suggests that they too are coming around to the same opinion.

In contrast to the influence of the minority Christian groups, the impact of the non Christian communities is more dramatic and often less peaceful. Head hunting by pagan tribes in Mountain Province attracts the greater publicity but it is the Moros, as the Moslem Filipinos are called, who have the greatest significance. The Moros have a radically different outlook on life to the average Filipino despite close racial and linguistic affinities. This stems from a fanatical adherence to Islam and its associated cultural patterns. The Moros even today still evoke fear on the part of the Christian Filipinos especially those from the Visayas who, despite Spanish protection, were continually harassed in previous centuries by the warlike Moros. The Moros are an independent and proud people. In the past they successfully resisted both the Spanish and the Japanese and, as the bloody confrontations between Moro and Christian in Lanao del Norte and Cotabato in 1971 demonstrated, they will continue to react in an equally hostile manner to both immigrants and government officials who attempt, through ignorance or design, to usurp their rightful place in the social and economic structure of Mindanao (and more particularly their traditional, though not necessarily legal, land rights).

Education. In a developing country such as the Philippines the strengths and weaknesses of its educational system can be of critical importance. Without an educated electorate, political power can be misused; without an educated farm community, land reform will not succeed nor can output be boosted by technological innovation; and, without education in the home, population growth can become intolerable. Fortunately, education is given a high priority in the Philippines: 30% of expenditure in the national budget is allocated to education. This in no small measure is a legacy, indeed probably the most important legacy, of the period of American control. American inspired free and universal public education does not in all respects match up to its American counterpart, but without it the social, economic and political betterment of the Filipino people would have been greatly retarded.

Throughout the Philippines, even in the most remote mountain regions, the opportunity exists to receive at least the rudiments of an elementary education. In extreme cases this may involve the pupils in long daily journeys over mountain tracks; they may even prefer to board with friends or relatives living close to the school. The latter state of affairs is more typical where the upper grades of primary education and more advanced schooling are concerned. For the average Filipino, however, primary and secondary education is readily accessible: four-fifths of all children aged seven to thirteen are in school. A high school diploma is an essential prerequisite to any ambitious Filipino; to attain it is as much a test of character as of educational prowess. This state of affairs arises from the shortcomings of the country's educational system. Despite, by Filipino standards, lavish government appropriations, pressures on the schools are such as to make overcrowding, inadequate equipment, poorly trained and overworked teachers the accepted state of affairs. With the school population doubling each decade or so, the level of financial support, however generous, cannot be expected adequately to meet even the minimum requirements.

Like many facets of the cultural environment, the basic educational system reveals no marked regional variations. Indeed, control of education is characterised by an exceptionally high degree of centralisation. Apart from a few cases of tertiary education, all educational activities are controlled and supervised by the Department of Education in Manila. Private schooling in the Philippines has arisen in part from the desire of the more wealthy members of society to secure a higher standard of education for their children and in part to capitalise upon the overcrowding characteristic of public education. The latter factor applies particularly to high school and vocational training. Many private schools in this sector, particularly those associated with the Catholic and other religious orders, perform a valuable service; others, however, may be just out and out diploma mills providing the minimum of instruction for the maximum fee.

Tertiary education is operated primarily on a private, profit making basis. The chief exception is to be found in Quezon City where the only state university, the University of the Philippines, is located. The bulk of the nation's undergraduates attend private institutions. Some of these, such as the University of the East in Manila, are huge (25 000-50 000 students), commercially operated institutions; others, such as the University of Sto. Tomas in Manila (founded 1611) and Silliman University at Dumaguete, are operated by religious or other essentially philanthropic organisations, emphasise quality rather than quantity, and make a notable, albeit restricted, impact upon cultural standards, not only on a nationwide basis but also upon the community which exists in their immediate environs.

Law and Order. An important factor, both in regional and national terms, affecting the life and livelihood of the Filipino is the state of law and order. Over most of the archipelago there is at least a tacit respect for and impartial encorcement of, law and order. To a stranger, and more especially a foreigner travelling in the country, the safety of his person and property can generally be guaranteed. Similarly, in many rural areas the work of the local police is confined to the occasional *bolo* fight resulting from the slighting of the Filipino's often highly sensitive pride, an over-indulgence of *tuba*, or petty larceny. On the other side of the coin, lawlessness, in all its forms, is rampant in many urban areas, corruption is widespread, and organised crime is able on occasion seriously to disrupt the life of the nation.

On a regional basis, law enforcement difficulties vary from the problem of forestalling the occasional forays by the fast disappearing headhunters of Northern Luzon to the control of the Hukbalahap (Huk) uprising. Whereas the headhunters of the north, and the Moros of the south, may cause terror as a result of their religious beliefs, the Huks are motivated by political and economic factors. The Huks were originally a wartime guerrila movement which gained communist overtones in the post-war years. Their policies, activity and strength have fluctuated over the years and it is now generally agreed that socio-economic reform, not force of arms, alone will secure a final solution. The Huks operate primarily in the Central Plain of Luzon (see Chapter Eight also) and are strongest in areas of exceptionally high tenancy, characterised by intolerable rents and abysmally low living standards. Despite several well intentioned efforts at land reform much remains to be done; until then, peasant discontent periodically will be exploited by agitators (likewise, the spectre of the Huks periodically will be raised by the politicians to divert the voter's attention from more pressing problems).

Perhaps more damaging than the menace of the Huks in recent years has been smuggling. Illicit entry of aliens, notably Chinese, and high value cargo, notably cigarettes, pose major problems for Customs officials. This is especially the case in the area of the Sulu Sea, whose island specked wastes and sparsely populated coasts enable the Moros, and their neighbours in Sabah, to successfully emulate the piratical trading propensities of their forefathers to whom international boundaries had no meaning. The profits to be derived from smuggling soon attracted big time operators who, aided by corruption in the Customs service, and in more prominent areas of government, were able to operate almost openly even in Manila Bay itself. In the first six months of 1968 alone contraband worth £16 million was seized during an anti-smuggling campaign.

Tax evasion also is prevalent, being on such a scale that the Bureau of Internal Revenue was forced in 1967 to initiate a heavy promotional campaign urging people to pay their taxes. As a result, the number of persons filling tax returns in 1968 topped the one million mark, representing a change of heart by more than 700 000 persons on the previous year. Until then, it had been estimated that there were more television owners than taxpayers. Tax evasion is not confined to individuals: in a recent case six cigarette firms were revealed to have avoided, with the assistance of corrupt tax officials, annual payments of £11 million by passing off as Virginia leaf tobacco, Native leaf tobacco chemically treated to look like a Virginia blend.

The open defiance of the law as practised by the smugglers has highlighted the most chronic of the problems of law enforcement—corruption. Essentially politically motivated, corruption to a greater or lesser degree is regarded as a perfectly natural phenomenon from the *poblacion* major upwards. As Anderson rightly points out:[4]

'Corruption is an ethical question; though some cosmopolitan Filipinos do define as 'corrupt' many of the social and political manipulations in which they indulge, most do not define them as such, as is shown by the fact that they are not secretive and feel neither guilt nor shame for such manipulations.'

In the past various efforts have been made to rid the country of corruption but not until the cost of smuggling reached such startling proportions did public opinion, as opposed to newspaper opinion, assert itself sufficiently to convince the politicians that an anti-corruption policy was worth more votes than could be bought through corruption. It is difficult to predict whether the current campaigns will be successful but if past experience is any guide their effects will not outlive them for any length of time.

Whether or not corruption is successfully tackled, violence appears destined to remain a permanent feature of Filipino society. Currently, the nation's murder rate is six times greater than that more well-known violent society—the United States. It needs to be stressed, however, that violence is very much an urban problem or, more precisely, a metropolitan problem. This in turn reflects the presence of organised crime, bad living conditions, unemployment, etc. Thus, Manila experiences an average of about 100 grave and less grave crimes per 10 000 population annually, but the comparable figure for the rest of the country is less than 2·5.

The combination of Malay and Spanish temperaments has produced a very volatile mixture in the Filipino. Easily provoked and slow to forgive, these weaknesses, which mar an otherwise likeable personality, have been exploited by politicians and criminals alike. The Huk uprising, organised crime Chicago-style, and pre-election gunning down of candidates are all symptomatic of the feature of Filipino heritage which more than anything else degrades both the way of life and standard of living.

Government and Politics. The constitution of the Republic of the Philippines was adopted in 1935. It bears a close similarity to that of the United States, which is not surprising as it was prepared under American guidance and required ratification by the United States Congress. The Constitution provides for a president, a cabinet of ten departmental heads, a Senate and House of Representatives and a Supreme Court, one of the few non-political, and hence impartial, institutes in the Philippine government. The president, elected on a four yearly basis with a maximum of two such terms, has vast executive and political power. Executive action has at times been hampered by conflict between the president and his congress but the president generally is in a commanding position as he holds the power of appointment for a large number of posts, both large and small.

Despite the great power accorded to the incumbent president, until 1969 none in the post-war years has been re-elected for a second term. The resultant lack of continuity in policy is exemplified by the fact that no less than a dozen economic plans have been formulated since independence but none has received the approval of Congress. Exploitation of his position and associated corruption have generally provided sufficient ammunition for the opposition to bring about a president's downfall. The two party system—Nationalists and Liberals—prevents any one party from dominating the scene. However, the apparent rigidity of the two party system is in some respects illusory. Both parties are highly personalised and followers frequently switch from one to the other, especially as their policies are indistinguishable in many respects.

The ultra democratic nature of Philippine politics so far has enabled the country to avoid the pitfalls created by the emergence of a personality such as Indonesia's Sukarno: on the other hand, leaders with the enduring popularity and maturity of Malaysia's Tunku Abdul Rahman are equally absent. At the regional level, however, local strongmen are not uncommon from the position of State Governor downwards. This is best exemplified by the Moslem areas where, following historical precedent, local *datus* retain in part at least their former feudal power, albeit now by money rather than force of arms. Alternatively, political strength may be proportionate to the personal fortune available. Vice President Lopez and his family, for example, not only own a chain of newspapers and television stations but also the country's largest electric power production unit; such is the wealth generated by these and other enterprises that the family recently made a bid to acquire the interests of the American Caltex Oil Company in the Philippines, including its oil refinery.

On balance, the quality of leadership in the Philippines is not as low as the behaviour of its political figureheads might suggest. Change in the Philippine Republic is generally for the better and, more important, is usually arrived at by peaceable and democratic methods. For this the ever growing core of well qualified and well trained (often in American universities) civil servants and technocrats, to be found not only in Manila but also in most provincial centres, must be held largely responsible. As a result, overall improvements to the Filipino way of life and livelihood since independence have been largely continuous though not spectacular. Future prospects, too, give rise to moderate optimism.

Architecture. The artistic character of a people has but limited geographical relevance; however, an exception needs to be made where architecture is concerned. Architectural styles contribute much to settlement geography, particularly where the urban scene is concerned. In the Philippines, architectural styles lack originality, with the exception of modifications caused by the employment of local materials. This state of affairs has arisen because building design has been imported either recently, from North America, or in the past, from Spain and, further back, from Indonesia and Malaysia. Only in the past few years have Filipino architects ventured to experiment, stimulated in part by the drive for cultural awareness sponsored by the wife of President Marcos.

The traditional Filipino house is a flimsy bamboo frame structure set off the ground on hardwood pilings (nearly two-thirds of all dwellings in 1970 fell into this category: of the 3 million structures involved one third were classified as substandard). Its Malay origins are self-evident. Regional styling usually reflects differences in available raw materials—bamboo, planks, *nipa, cogon,* etc. The Spanish introduced a very different form of architecture characterised by massive stone construction at ground level topped by a wood and tile second storey. Formal ornate structures were typical and introduced a very alien form to the urban scene hitherto little different in appearance from that of neighbouring South-East Asian nations. Spanish styles were, however, incorporated in the traditional structure where the occupier could afford it, as in the employment of decorative ironwork and the use of stone masonry instead of hardwood piling. In the twentieth century modern building materials such as galvanised iron and glass windows have further modified building styles while rising living standards have transformed interiors (however, in 1970 only 20% of dwellings possessed piped water and only 16% contained a bath and toilet). Then, after World War II, domestic architecture became strongly influenced by America—bungalow/ranch style dwellings with formal gardens are increasingly common in the better suburbs, and within, consumer durables and household gadgets increasingly proliferate. Likewise in the major urban centres, office blocks, factories, etc., bear a close resemblance to those to be found in North America.

Traditional styles predominate to the virtual exclusion of more recent variants beyond the bounds of the principal urban centres. In contrast, modern styles of public architecture in the forms of churches, schools, town halls, etc., are distributed rather more widely. Nevertheless, an urban concentration is again evident. Spanish public architecture favoured massive stone construction, a factor that enabled the survival of many a church, for example, in disturbed parts of the country. Non-residential building increased sharply in the American era and styles reflected this new source of architectural ideas. More recently, Filipino-inspired styles have come to the fore. These take greater account of local climatic conditions and building materials whilst attempting to retain traditional styles.

Population

Various demographic features of the Philippines require close study by the geographer. The ramifications of population growth, for example, involve consideration of inter-regional movements, urbanisation, and pioneer settlement. In the context of Philippine socio-economic development the manpower resources available need to be viewed in the light of job opportunities, particularly as regards under- and unemployment, and of Filipinisation of employment sectors, particularly in relation to the emergence of a viable middle class stratum of society. Perhaps most important is an appreciation of the dynamic nature of the Philippine population, not just as regards the future implications of the slowness in introducing of birth control measures in a predominantly Roman Catholic society but also, for example, the impact of the current youthfulness of the population and the exhaustion of new areas for rural resettlement.

In the following pages are reviewed some of the country's outstanding demographic features. Statistics are derived either

TABLE 18

Population, Land Area, % Change, Density and % Distribution, 1960-70

	Population ('000 persons)		Land Area		Population Change		Density		% of Total Population	
	1970	1960	sq. km	%	1960-70	1948-60	1970	1960	1970	1960
Philippines	37 088	27 008	300 000	100	36.6	40.8	123.4	90.3	100.0	100.0
Manila	1311	1139	38	neg.	15.1	15.7	34 219.4	29 728.7	3.5	4.2
Ilocos and Mt. Province	1846	1470	25 766	8.6	25.6	30.1	71.1	57.0	5.0	5.4
Cagayan Valley	1478	1036	26 838	9.0	42.7	54.8	55.1	38.6	4.0	3.8
Central Luzon	5099	3691	23 646	7.9	38.2	33.0	215.7	156.1	13.8	13.6
Southern Luzon[a]	6767	4232	46 119	15.4	59.9	66.4	146.7	91.8	18.3	15.6
Bicol	2996	2363	17 633	5.9	26.8	41.8	169.9	134.0	8.1	8.7
Western Visayas	3873	3210	21 579	7.2	20.7	21.6	179.5	148.8	10.5	11.9
Eastern Visayas	5390	4564	36 383	12.1	18.1	17.5	148.1	125.4	14.6	16.9
Northern Mindanao	3332	2111	39 845	13.3	57.8	53.4	83.6	53.0	9.0	7.8
Southern Mindanao	4916	3273	62 153	20.7	50.2	108.8	79.1	52.7	13.3	12.1

Source: Bureau of Census and Statistics.
[a] Including Mindoro and Palawan

from Census material or from those collected by means of the Bureau of Census and Statistics' *Surveys of Households*. Detailed discussion of their implications for the geography of the Philippines, however, are in general deferred until the concluding chapters, where they can be viewed in their proper perspective.

Basic Characteristics. The 1970 population of the Philippines has been estimated at 37·0 million, compared with the 1960 Census total of 27 087 685 (Table 18). The 1970 figure is the product of a 3·2-3·4% annual growth rate (between 1948 and 1960 the population increased at 3·18%). The Republic's population thus ranks amongst the fastest growing in the world. However, by Asian standards, in themselves above average, the population density of the Philippines, at 123 persons per square kilometre in 1970, is high but not excessively so: in 1968 the comparable figure was 120 persons per square kilometre (Table 19) and the nutritional density was 358 persons but at that time Taiwan recorded comparable figures of 372 and 1535 respectively. But, with the population total expected to double in the next generation, the heavily populated areas now characteristic of the island of Cebu and parts of Luzon could become a major feature of the pattern of occupation.

The nation's population growth is caused by a fall in the death rate rather than an increase in the birth rate. The recorded death rate (the real rate is perhaps double the official one) dropped from 16.5 persons per 1000 in 1939 to 7 persons per 1000 in 1960 (a level apparently maintained in 1970). The unofficial birth rate has consistently stayed in the vicinity of 45-50 persons per 1000[5]. Universal and generally adequate medical facilities and techniques, established in the early part of the century by the Americans, are largely responsible for the fall in the death rate. Infant mortality in particular has been cut drastically while malarial control has prolonged many lives also. Today, tuberculosis is the chief mortality problem and if, as is probable, this disease is brought under control, a further upsurge in the population growth rate could result.

The lack of significant change in the birth rate of the Philippine people is closely tied in with religious beliefs. The majority of the population professes the Roman Catholic faith and, furthermore, remains obedient to the dictates of an

TABLE 19
Area, Population and Density of Selected Asian Countries[a]

	Population	Total Area (sq. km)	Gross Population Density (sq. km)	Cultivated Area (sq. km)	Nutritional Density (sq. km)[b]
Malaysia	10 034 000	332 633	31	33 240	310
Burma	25 246 000	676 578	38	160 880	151
Thailand	30 744 000	514 000	60	114 150	269
Indonesia	92 600 000	1 491 564	62	176 900	525
China (Mainland)	646 530 000	9 561 000	68	1 093 540	595
South Viet Nam	17 414 000	173 809	100	28 370	614
PHILIPPINES	35 883 000	300 000	120	100 270	358
Pakistan	120 160 000	946 716	127	284 290	423
India	511 125 000	3 045 220	168	1 583 410	323
Taiwan	13 466 000	36 151	372	8 790	1535

Source: Based on R. E. Huke, *Shadows on the Land: An Economic Geography of the Philippines*, Bookmark Inc., Manila, 1963. Updated by reference to ECAFE data.

[a] Data all relate to the latest available year for cultivated area in each country. This is 1967 or 1968 except for Indonesia and Mainland China (1960), Thailand (1965) and Burma (1966).

[b] Gross population per sq. km of gross cropped area (double cropped areas counted twice). Use of this index improves the position of the Philippines vis à vis other Asian nations as the proportion of gross cropped area is comparatively high.

ecclesiastical hierarchy noted for its traditionalism. The resultant rarity of birth control measures means that in the foreseeable future the country will have to live with the frustration of seeing the rewards in economic advancement enjoyed not by their creators but diverted into measures to ensure that by running as fast as it can the country remains in the same place. Japan's success in stabilising its population growth to the stage whereby, Western-style, two children families are commonplace, now must be regarded as an impossible goal for the Philippines in the twentieth century; more disturbing, however, is the increasing likelihood that the Philippines equally will not come to emulate Taiwan which took the first step along the same road as Japan in the last decade by reducing its birth rate by one third, from 42 to 28 per 1000 persons.

The fall in the death rate, coupled with increasing urbanisation, unfortunately may have elevated the health hazard of malnutrition into a major long term problem. Some observers fear that malnutrition in pre-school children produces permanent mental retardation. The long term implications for a country such as the Philippines with its exceptionally youthful population can well be imagined. More specifically, in Manila's chief slum area, Tondo, recent work by the Catholic Relief Services suggests that one out of every six pre-school children is suffering from severe malnutrition. Since in some Philippine cities up to 45% of the population exist in slum conditions, not to mention many others living in rural slums, there could be at present a quarter of a million pre-school children suffering from severe malnutrition, plus a further million or more suffering from moderate malnutrition. The resultant waste of human resources is one that the Philippines can ill afford yet will have to live with for at least the next generation.

The success achieved in respect of death rates but the failure in respect of birth rates has produced an exceptionally youthful population in the Philippines. Currently, the median age stands at about seventeen years; in 1903 it was over twenty years. Preliminary results of the 1970 Census suggest that 57% of the population was below twenty years of age; by 1980, therefore, the numbers in this category are expected to have risen to a figure greater than that of the total 1960 population. This situation means that, even in the unlikely event of an early and significant fall in the birth rate due to the introduction of birth control measures, substantial growth in total population is inevitable into the twenty-first century. This situation arises because there is a twenty—thirty year lag between the inception of a fall in the birth rate and a fall in reproduction rates. And with a youthful population it is the latter which is the key element in population growth. Currently in the Philippines the incidence of child bearing, expressed as the ratio of children below five years old for every 100 women aged fifteen to forty-four, is 95.31 (in other words, on average every mother possesses one infant child). Families may become smaller through a combination of Filipino women participating more actively in the work force and servant labour becoming more expensive. But, until Roman Catholic attitudes to birth control are modified, it will remain a quite commonplace state of affairs for Filipino women to bear seven or more children, almost all of whom now reach maturity.

Urban life appears in general to have a beneficial influence on the population problems of the Philippines. Greater leisure opportunities lead to greater sophistication, delayed marriages and ultimately to smaller, albeit only slightly smaller, families. It will be in the urban areas, too, that any revolt against current Roman Catholic birth control doctrine will be initiated. Statis-

TABLE 20

Selected Statistics of Young People of the Philippines

| | Population of Young People 0-24 years | | Est. Young People per 100 working pop. (1969) | |
	Est. Geometric Rate of Growth (1960-69)	Index	Nos.	Index
Philippines	3.60	100	54.67	100
City of Manila	3.53	98	43.67	80
Ilocos and Mt. Province	3.75	104	50.86	93
Cagayan Valley	3.64	101	50.89	101
Central Luzon	3.61	100	54.93	100
Southern Luzon	3.67	102	54.00	100
Bicol and Masbate	3.51	98	57.35	100
Western Visayas	3.55	99	55.02	100
Eastern Visayas	3.55	99	54.15	100
Northern Mindanao	3.61	100	56.60	100
Southern Mindanao	3.60	100	57.11	100

Source: Anon, 'Children and Youth Population of the Philippines,' *Journal of Philippine Statistics,* Vol. 19 No. 4, Oct.-Dec. 1968, pp. ix-xii B.

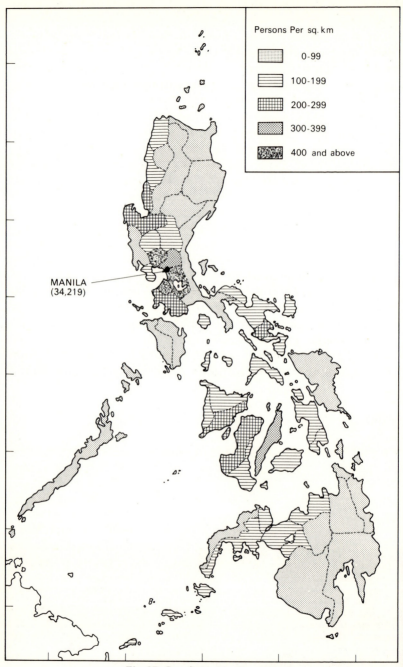

Fig. 27. Population Density, 1970

tics can be misleading due to the fact that the urban population, being the product of natural growth plus inmigration, possesses an age structure distorted by a relative absence of older people and a sex structure distorted by a relative surplus of young males. Nevertheless, it would appear significant that the geometric rate of population growth of persons aged 0-24 years for Manila, at 3·53%, is lower than comparable rates for all other parts of the country except Bicol (Table 20). Additionally, Manila possesses easily the lowest child dependency load, at 43·67 children per 100 working adults compared with a national average of 54·67 children per 100 working adults.

Distribution. As already suggested the overall population density is subject to considerable areal variations. Latest figures reveal that Rizal Province possesses the highest gross density (excluding Manila proper), with 1495·5 persons per square kilometre (Fig. 27). Net, or physiologic, densities are much higher and figures in excess of 5000 persons per square kilometre of cultivated land occur in several provinces. The high gross densities in Rizal and adjacent provinces such as Cavite (402.7 persons) and Laguna (408.3 persons) reflect the ever increasing urbanisation of the metropolis and its environs (comparable densities in 1960 were 293.7 and 268.3 persons). High population densities are typical of most of the Central Plain of Luzon and also of the more intensively cultivated areas of the Visayas and the Bicol Peninsula. The empty areas of the country are dominated by the island of Mindanao but in addition Palawan, Mindoro and Northern Luzon are strikingly devoid of people. Palawan, with 15.6 persons per square kilometre, is the most sparsely populated province of the Philippines.

When rates of population growth are examined a contrasting picture emerges (Fig. 28). In this case, much of Mindanao, Palawan and Northern Luzon join already densely populated areas such as Rizal Province in exhibiting above average population growth rates in the period 1960-70. Growth rates in excess of 75% for this period (the national average being 36·6%) are recorded by three provinces: one is Rizal and the other two (Lanao del Sur and Bukidnon) are located in Mindanao, with Bukidnon recording the highest growth rate in the country (106%). Below average rates of growth are generally character-

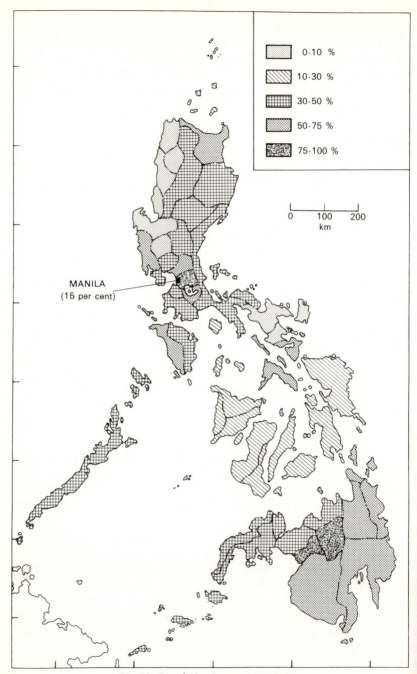

Fig. 28. Population Increase, 1960–70

istic of areas where pressure on available farm land is already high and few alternative employment opportunities are manifest. The Visayan provinces are typical examples of this phenomenon, characteristic growth rates being shown by Bohol (14%) and Samar (15%), although the least impressive growth rate (4%) was recorded by the nearby province of Catanduanes in South-Eastern Luzon.

Mobility. The great regional disparities in population growth indicate that the population possesses a high degree of mobility. This is not surprising for the Philippines was peopled in only the relatively recent past and, furthermore, being an end migration area is still subject to major readjustment of its population pattern. Thus, the bulk of population movement is internal in nature: immigration is of negligible proportions, especially now that there is strict control on the entry of Chinese; emigration is at a somewhat higher level, being mainly to the United States, but here too levels are lower than in the past.

Internal migration takes three forms. The most spectacular is seasonal in nature. Then, there is migration connected with urbanisation which is most evident in the case of Manila. The third form of internal migration is the movement to the 'pioneer lands' of which the island of Mindanao is the supreme example.

Seasonal movements related to the sugar industry overshadow all others. During the milling season the unemployment and underemployment characteristic of the sugarlands disappears overnight. For over five months either side of the New Year, labour becomes a valuable commodity but its value is depressed because the labour shortage attracts people from far and wide. Thus, in Negros Occidental the number of resident workers is swollen 60% by some 30 000 temporary immigrants largely from the neighbouring island of Panay which, despite possessing sugarlands of its own, still possesses a surplus labour force drawn mainly from the less fertile western provinces of Antique and Aklan. Paralleling the situation in Negros, workers are recruited for the sugar milling season in Luzon from the overpopulated Ilocos coast. This involves greater distances, perhaps 200 kilometres, for some, but the timing of the milling season is attractive in that it largely coincides with the dry season when agricultural activity reaches its low point. A third

TABLE 21
Population Growth of Selected Urban Areas

	Population Total, 1970[a] ('000 persons)	Increase over 1960[a] ('000 persons)	(%)	Increase, 1948-60 (%)
1. Metropolitan Manila	3000	892	44.2	55.9[c]
Manila Proper	1311	172	15.1	16.4
Quezon City	575	177	46.7	268.0
Caloocan City	175	55	45.8	145.3
Pasay City	184	51	38.3	49.0
Other Areas	755[b]	437	139.6	n.a.
2. Other Principal Urban Centres[f]	1120	326	40.8	51.1[e]
Cebu	325	100	44.4	49.9
Iloilo	213	63	42.0	37.3
Davao	125	40	47.1	102.9
Bacolod	125	32	34.4	17.6
Angeles	109	33	43.4	n.a.
Cavite	77	22	40.0	56.6
Legaspi	75	15	25.0	−22.8[d]
Baguio	71	21	42.0	72.4
3. Total Population	37 008	9920	36.6	40.8

Source: Bureau of Census and Statistics.

a Based on Table 2.
b Estimate of suburban population of Makati, Mandaluyong, Parañaque, San Juan and other parts of Rizal and Bulacan.
c Excludes parts of Rizal and Bulacan.
d Result of boundary change.
e All Chartered Cities.
f With over 70 000 persons in 1970

internal migration of considerable proportions is that from the Central Plain of Luzon and the Ilocos coast to the neighbouring Cagayan Valley to assist in maize planting and harvesting. This movement occurs in the period June-September and thus does not attract those who are required for essential agricultural work in their home area.

A seasonal movement of population of a different nature occurs in the second half of December. This is the time when the urban dwellers seek to return to celebrate Christmas and the New Year with the relatives and friends left behind in the home *poblacion, barrio* or *sitio*. There are other former residents who have set out to seek a better way of life elsewhere in the archipelago but the urban dweller predominates amongst those possessing the means to make such a journey. They, too, represent only a fraction of those who have migrated to the towns and cities, for many merely swell the numbers of slum dwellers. Unlike the agriculture oriented seasonal migrations, that at Christmas benefits all modes of transportation, not just the overcrowded ferry boat or inter-provincial bus.

Migration connected with urbanisation has its origin in every corner of the Republic but one destination is preeminent—Manila. The governmental and industrial employment opportunities of the metropolis make it a mecca in the eyes of rural inhabitants for whom unemployment and underemployment are a way of life, and for workers in other urban centres who, having experienced the more sophisticated way of urban life, are attracted by the range of facilities to be found in Manila. Such is the migration to the metropolis that its annual population growth rate in the period 1948-60 was one third greater, in some sections six times greater, than the national average; between 1960 and 1970 the metropolitan growth rate was roughly one fifth greater (Table 21). The consequences of this inflow are examined in more detail in Chapters Eight and Nine.

Whereas movements to the cities still are gaining in momentum, movements to the pioneer areas of the country face an uncertain future. Most pioneers originate in the Visayas because the inhabitants of Luzon exhibit but limited mobility in this sphere. The Visayan, often being in more intimate contact with the sea and, in the case of the Cebuano, facing severe population pressure, is more appreciative of, and able to cope with, the challenge of pioneering. In Luzon the Ilocanos are a major

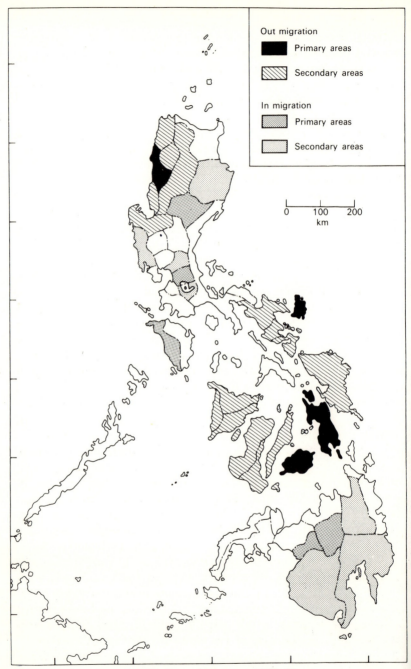

Fig. 29. Internal Migration, 1960–70

exception, facing as they do similar pressures as the Cebuano. Nevertheless, they prefer the adjacent Cagayan Valley as a place of new settlement rather than the principal target of the pioneer—Mindanao (Fig. 29).

In the inter-war years Mindanao was regarded in many quarters of the American Administration as a potential Texas but greater knowledge of the island has brought the realisation that this was an overoptimistic assessment. Nonetheless, in the period 1948—60, the island's population was able to be virtually doubled to reach 5 million persons, and a further 1·5 million was added to this total in the next seven years. By that date each person occupied on average 1·47 hectares of surface area compared with an equivalent figure for 1913 of ten hectares. Unfortunately, more than half the surface area of Mindanao is unsuitable for agricultural activities with the result that current settlement has already appropriated nearly two-thirds of all suitable land. The potential of the island for new settlement has thus been reduced to a level which is well below the total annual population increase for the whole country. If current trends persist therefore, there will be little new land available for agricultural settlement in Mindanao by the mid-1970s. With similar pressure building up in other areas, the pioneer frontier in the Philippines will have been pushed back to its ultimate limit by the late 1970s.

Employment. It has been demonstrated already that the youthful nature of the Philippine population implies a very high dependency ratio. This would present but limited problems in a situation of high incomes and full employment but the reverse is the case where the Philippines is concerned. The high dependency ratio places a very heavy burden upon adult members of a family, for often one or more of its members will be unemployed and when work is available its returns often are meagre. The close kinship ties characteristic of Filipino families helps to spread the burden but, with unemployment standing at about 7-8% and with underemployment an estimated 25%, income sharing has its limits.

The total labour force of the Philippines in 1970 just exceeded 13 million persons. Of this total only about 9 million were in full-time employment. In the agricultural sector, the principal source of employment, are to be found three-quarters of all self-employed workers and nine-tenths of all unpaid family workers. Here lies the source of most underemployment. Many people in this category, particularly unpaid family workers, have no other choice but to participate in a family landholding which, because of insufficient land and capital, can produce only meagre returns. A parallel situation is to be found in urban areas, particularly in cottage industries and retail trade. However, it should not be forgotten that a substantial proportion, slightly over half in the case of agriculture, of part time workers (whether paid or unpaid) were unable, because of other commitments, e.g. school or housework, or age and infirmity, to engage in full time employment. Nevertheless, in agriculture alone this leaves a residue of almost one million persons forced by circumstances beyond their control, to work for a shorter period of time than they desired.

The incidence of unemployment probably is greater in urban areas. As such it differs from underemployment. Unemployment has many causes but probably outstanding in the Philippines is the fact that the inflow of people from the countryside to the towns exceeds the creation of job opportunities there. Furthermore, newcomers to the urban areas do not always have recourse to relatives and friends who can provide partial employment with the result that the incidence of concealed unemployment is below average. The resultant inability of the individual to 'save face' in respect of employment exacerbates his frustration and contributes to the law and order problems facing urban authorities.

The preponderance of people in the rural areas of the Philippines coupled with the greater incidence of underemployment there means that rural development is the key to the containment of population pressure in the Philippines. Stimulation and improvement of the rural labour force is retarded because of its conservativism and ignorance. Rural inhabitants overall are less skilled, operate with lower efficiency, receive lower incomes and save less than their urban counterparts. Improvements to these sectors cannot be achieved until the surplus labour force is eliminated, for it is this more than anything else which depresses the earning ability of the rural inhabitants. There appears little scope to eliminate the existing surplus, and that to be created by ever increasing population pressure in the future, by creating more rural employment opportunities. Here, the dilemma is whether to increase the earning ability of a select few through the injection of capital and technical know-how thus increasing the size of the surplus work force, or whether to increase rural output through the medium of existing labour-intensive methods, thus diluting the impact of increased earnings. Given that rural development implies the location of employment opportunities in non-rural occupations, it appears natural that industrialisation offers a way out. But indiscriminate industrialisation would have only short term benefit, hence the government's policy of preferential treatment for basic manufacturing and related activities. In the longer term, manufacturing, with its increasing need to achieve efficiency through capital-intensive operations, will prove inadequate to absorb the ever increasing work force. Then, it will be the turn of the tertiary service sector to bear the full brunt of job seekers with a probable proliferation of unproductive employment outlets. The Philippines then will not only be running as fast as it can to keep in the same place but it will be running in a pointless direction.

Land Tenure

The land tenure pattern in the Philippines is founded on share-crop tenancy, one of the least desirable arrangements devised between the person who owns the land and the person who works the land. Agrarian reform thus is closely bound up with land tenure reform but success in this sphere, like the containment of population pressure, so far has had limited success, nor are prospects for the future much better. The reasons for this state of affairs are examined in detail in Chapters Eight and Nine; the background to it and other characteristics of land tenure forms the subject of the present discussion, as such phenomena can have an important bearing on a range of features of interest to the geographer such as land use patterns, crop productivity, rural settlement patterns.

In the Philippines the best lands are in the hands of a few owners,[6] leaving the majority of landholders with only small plots often on marginal or sub-marginal lands. The pattern of urban landownership parallels that of rural areas; most choice sites are in the possession of a few wealthy individuals, leaving the would-be small businessman or householder little choice but to accept a tenancy status or else content himself with an inferior location. The debilitating influence of inequitable land ownership also enables excessive political and commercial power to be exerted by the principal landowners.[7] This can result in the mass of the inhabitants being conditioned into a passive acceptance of the existing socio-economic structure; even in the more educated Filipino, the desire for leadership is muted; when major decisions are to be made most prefer to follow rather than lead and to allow someone else to do their thinking.[8]

Existing land tenure systems cause half the nation's farmers to be in a state of perpetual indebtedness. Improved farm practices do little to alleviate the situation, for the bulk of the net proceeds resulting are received by the landlord through the terms of the tenancy agreement. Tenancy in rural areas typically takes the form either of share tenancy (the *kasama* system) or leasehold tenancy (the *inquilinato* system). Share tenancy is the more common although officially illegal. It involves the furnishing of labour and work animals by the tenant while the landlord supplies the land, farm implements, seeds and expenses for planting and harvesting labourers. Usually the tenant receives the greater, but not necessarily an equitable, portion of farm income. In contrast, leasehold tenancy consists of the tenant leasing the landlord's land for an annual rent. The tenant is thus largely independent as he finances the entire operation. He, however, bears all the risks for the amount of the rent (or *canon*) is not determined by the size of the crop but by the average potential of the land. For the leasehold tenant therefore the all too common crop failure can be a major disaster; for the share tenant the entire tenancy arrangement is usually a disaster. The latter finds himself confined by the often unjust share of production taken by the landlord and the low productivity of his land. Being often ignorant and conservative in outlook he is reluctant to seek advice, even if any were at hand, and usually is only able to preserve his meagre standard of living by borrowing at usurious rates. Once this happens his indebtness is permanent for no matter how hard he works his output, even in good years, offers little opportunity to amass a surplus sufficient to banish his insolvency.

The near-feudal situation under which many a farmer's way

of life and livelihood is formulated is not usually as simple as portrayed above. Numerous local variants exist and the situation is further complicated by official and unofficial subletting of the landholding in its entirety or in fragments. Generalisations in respect of land reform therefore are dangerous and the Philippines is fortunate that its basis for action in this sphere—the 1963 Agricultural Land Reform Code—reflects 'a highly sophisticated insight into the changes that must be associated with the land tenure modifications if the productivity potentials inherent in the legislation are to be achieved.'[9]

The objectives of the Agricultural Land Reform Code are:

1. To establish owner operated farms of economic family size as the basis of Philippine agriculture;
2. To free farmers from pernicious institutional restraints and practices;
3. To achieve greater productivity and higher farm incomes through the creation of a truly viable social and economic structure in agriculture;
4. To apply all labour laws to both industrial and agricultural wage earners;
5. To provide a more vigorous and systematic land resettlement programme and public land distribution;
6. To make small farmers more independent, self-reliant and responsible citizens.

The land tenure reform which is vital to the achievement of the above objectives realistically envisages a gradual restructuring of the existing system. Agricultural share tenancy has been declared contrary to the public interest and is to be abolished except where it relates to fishponds, saltbeds and land planted to permanent trees such as coconuts, citrus and durian. The share tenants first are to be shifted into leasehold. The fixed rental paid to the landlord is to be the equivalent of one quarter of the average normal harvest during the past three years, after deduction of the amount used for seed and the cost of harvesting and processing. The next step is to convert the lessees into owner operators which is achieved through the government purchasing the private agricultural lands worked by the lessees; in the period during which the former lessees are paying for the land allocated them by the government they are known as amortising owners.

Action on various fronts is envisaged to execute the operational aspects of the Code. Leasehold status, being an intermediate phase, involves appreciating the conditions of ownership and to assist in this a package of services is provided through the Land Reform Project Administration. A key element of the package is supervised credit. This approach is being paralleled in those areas which, prior to the establishment of the Code, new land settlement and government acquisition of private agricultural land already had been in progress for some time. The Land Bank has a vital role to play. Through this organisation former landlords receive in cash 10% of the value of the land acquired; the balance is credited in the form of interest earning bonds (these forced savings then are invested by the Bank in industrial development projects). Paralleling the acquisition of private land is a programme to open up public lands either to resettle farmers displaced as a result of the creation of economic family size farms or to make provision for former landlords who may wish to reinvest in agriculture. The latter, however, would be restricted to large scale commercial enterprises such as oil palm, beef cattle and fishpond farming. Finally, an intensive training programme has been initiated to ensure understanding and appreciation of the changes resulting from the implementation of the Code. Training is designed not only to create a responsible, knowledgeable and inspired corps of land reform personnel but also to ensure that the farmer's response is enlightened and positive.

If these laudable objectives are achieved the people of the Philippines will have secured an immense advance in social and economic development; but in recent years many well-intentioned attempts have been made at land reform and all have failed. For each the story is one of empty promises, resistance by vested interests, inefficient administration, apathy and ignorance on the part of sharecroppers, and similar discouragements. The Agricultural Land Reform Code is realistic enough to anticipate the problems of its implementation: in establishing a fair rent for the leasehold period a credible record of past harvests often is difficult to obtain; subdivision to establish family size units on acquired or public land is a protracted process due to the administrative and survey procedures required; landlords are not convinced of the stability of the Land Bank bonds; and the quality and quantity of staff available leaves much to be desired. These are but a few of the difficulties facing the various government organisations charged with implementing the Code: some, ironically enough, have their origin in other branches of government.

Anticipation of land reform problems does not ensure their solution. Progress to date by the Land Authority in achieving the objectives of the Code (Table 22) suggest that only 16% of

TABLE 22
Implementation of the Agricultural Land Reform Code, as at June 30, 1971

	Farm Families	Area (hectares)
Conversion of palay/maize share tenants to lessees[a]	182 203	408 278
Recipients of production loans[b]	101 937	327 438
Direct administration by the Land Authority	86 000	500 000

Source: Fookien Times Yearbook, 1971.

[a] Represents only totals for municipalities proclaimed as placed under leasehold; actual conversion probably is far from being completely effective. The number of farm families represents approximately 16% of all the nation's share tenants as defined by the Land Authority.

[b] In period 1964-70.

share tenants have been affected. It is too early yet to conclude that the Code will go the way of its predecessors but it has yet to arrest the deterioration which has occurred since the 1948 Census of Agriculture: indeed, some estimates suggest a subsequent 50% rise in the incidence of tenancy.[10] Any such deterioration of the situation would appear concentrated in those areas already possessing a major tenancy problem. Such areas are dominated by the sugar and rice lands: in Pampanga Province on the Central Plain of Luzon almost nine out of every ten farmers were tenants in 1948 while in sugar rich Negros Occidental the comparable figure was almost two out of every three farmers. Tenancy largely disappears in the least attractive agricultural areas; in 1948 only in remote Batanes Province, rugged Mountain Province and undeveloped Palawan Province did more than 90% of all farmers own the land they operated.

The concentration of land ownership is perpetuated today not only by inheritance but also by its social, economic and political benefits. In the Philippines, the possession of land and, by inference, its tenant population not only provides a sizeable and relatively troublefree source of income but also bestows upon the recipient considerable political power and hence social prestige. A desire to keep what land one has and to obtain more if possible is thus a major motivational force in Philippine society. It is one which is not confined to its upper echelons; the small landholder/merchant of the rural *poblacion* has a similar attitude to that of the wealthy *haciendero*. Land reform therefore must be concerned as much with changing people's attitudes as with changing the tenure of the land.

Social Attitudes

For all its past and present contact with and assimilation of Western culture patterns, the Philippines remains in essence an Oriental nation. One of the most important manifestations of this fact is to be found in people's attitudes towards one another. Such attitudes not only may contrast sharply with those characteristic of the West but, as will be demonstrated, make a major contribution to the way of life and livelihood in the Philippines.

Individual Attitudes. Despite the limited participation of women in economic life, a matriarchal society exists in the Philippines. Within the family it is the authority of the mother and not the father which is paramount, even in money matters; the womenfolk too play a major role in communal life and in political affairs (one sixth of the Senate in 1968 were women). Even in the economic sphere there is a significant amount of indirect participation for a son may often submit to his mother for approval of his choice of job or business approach.

A web of relationships is contained in any one family, reflecting its origin in the interrelated Malayo-Polynesian tribal or *barangay* system. In addition to his immediate relatives, the average Filipino's 'family' also comprises cousins to the nth degree, two sets of godparents (one for christening and one for confirmation), and various other satellite kin who may include business associates and neighbours. Such a centripetal system—to use an anthropological term—involves the mastery of a complicated system of behaviour ranging from handkissing as a mark of respect to one's elders to the courtship pattern.

Modern civilisation may have weakened 'family' ties in some instances but its influence remains very strong as the following example quoted by Anderson[11] illustrates:

'Tomas is a well-known young businessman in Manila. Nestor is the chief agent in the province of Cagayan of one of Tomas's enterprises, a logging concession. Nestor has a son who, at the time of this story, was a student in a Manila university. The boy had no close relatives in Manila and his parents were concerned about both his safety and his ability to meet necessary expenses. Too embarrassed to approach the boss directly, Nestor went to the home of Tomas's brother-in-law, who he had previously served as a trusted employee and who had recommended him for his present job. He raised his worries concerning his son and his fears that the son might fall into trouble without proper supervision. Later, the brother-in-law approached Tomas regarding the matter. Tomas replied that he would be happy to help and raised the topic at his next meeting with Nestor. He not only provided part-time employment for the boy in his Manila office but also took him under his supervision, advising him to open a savings account for his excess earning, to score high grades, and to avoid association with companions who might lead him into trouble.'

In many instances *Ugaling Pilipino*, as family behaviour patterns are known, has little external impact; it, however, significantly influences the life and livelihood of the community, with politics outstanding in this respect. The size and complexity of the family often causes the eldest child to act as a third parent and such duties can be sufficiently onerous to affect one's job or postpone marriage. The development of a family compound to house the multiple relations adds a unique element to the settlement pattern, while inclinations towards migration may often be frustrated by the social and moral obligations to the rest of the family. On the other hand, the intertwining of family relationships makes possible job opportunities, secured at times by often exceptional hardships on the part of parents, contact between widely differing social classes, and moral and economic support in times of difficulty. *Ugaling Pilipino* has important repercussions in economic activity. As Anderson[12] points out, status seeking and consciousness of rank, for example, are strong motive forces which result in multiple employment at all levels of society. Thus, a high school principal operates a small transportation company and acts as a tobacco buyer; a dock labourer may also work for a fish merchant and repair roads. Multiple employment is especially characteristic of trade; middlemen abound, most having gained initial experience selling sweepstake tickets, *sampaguita* flowers, religious medals, or more prosaic items such as cigarettes, fruit or ice cream. All levels of society may act as agents during business hours: friends of employees bring jewelry, blouses, bags, etc.; during breaks offices are invaded by tailors, caterers, etc.

Such middlemen depend upon a regular clientele (for which the family provides the obvious starting point) with whom they maintain personal relationships permitting sales on short term credit. This relationship is often the stepping stone to larger scale business enterprises but the social *mores* persist with considerable strength. Economic activity, and particularly commerce, is strongly coloured by these constraints, notably the *suki* relationship whereby a stranger is made 'friend' and sharp practices are replaced by concern for personal satisfaction and loyalty. In a developing economy such as the Philippines, credit is a vital form of finance and mutual trust; business can only be developed by means of social obligations—the risks attendant to impersonal business relationships make sharp practice an inescapable way of life.

An example quoted by Anderson[13] will serve more clearly to illustrate the situation:

'Mr. and Mrs. M. are full-time *commerciantes* who deal in a wide range of goods. They own a small commercial fishpond (which they let to a tenant, since their trading activities require too much mobility for them to tend their holding). Early each day they set out by bus for Reina Maria with fresh fish, crabs, shrimp, pork or carabao meat, and *bagoong* (salted fish sauce). There they rent a carriage, and on consecutive days they visit large barrios that line the roads to the north-east and south-east. These barrios constitute their 'territory', where they have sold for years and have established *suki* buyers. Their prices are a bit higher than those haggled in the public market of the municipality, but their *suki*, who would have to pay transportation fare if they went to market, reason that the additional cost is easily outweighted by 'home service,' consistently high quality, and extension of credit. Credit, at no interest, and willingness to engage in barter are particularly valued given the relative shortage of cash in the barrio. Like all buyers-and-sellers the M.s constantly work toward maintaining a workable level of liquid assets, limiting their extension of credit. Under conditions of heavy supply, engaging in barter in agricultural products which are readily saleable conserves cash and provides favorable exchange rates when the products are sold in their hometown (where such agricultural produce is limited). Thus, bartered products gain in value just from the trip, as do the fish products. Mr. and Mrs. M. procure their stocks in three ways: from their own production; by bulk purchase (on credit, payable after stocks are sold), from kinsmen or *suki* who are emptying and restocking a fish pond; or by purchase from a middleman *suki* who sells large stocks in their town market. The second procedure is the most common. *Suki* linked in personal alliances reciprocate the advancement of stocks to one another as each fishpond is harvested.'

The above social attitudes have been discussed at length, for in a developing country such as the Philippines commerce is a vital key to growth and its efficiency depends more on entrepreneurial skill than technical innovation. As Anderson[14] puts it:

'Filipino entrepreneurs are the key men who, in the absence of a tradition which includes corporate organisational forms or joint, horizontally integrated, economic enterprises, form around themselves a highly flexible personal coalition combining necessary elements of knowledge, authority, skill, capital, labour, markets, and influence ... The lack of real opportunities, the relatively low rate of success, and the sizeable rate of failure do not represent serious challenges. Buy-and-sell and other small scale businesses appear to create part of the reservoir of talent which can expand operations and from which entrepreneurs may be drawn ... economic development in a developing non-socialist economy may depend less upon a few large scale entrepreneurs and more upon widespread if small scale entrepreneurial activity. Large scale enterprise becomes of overwhelming importance later in the process. Finally, despite the obvious and oft-cited maladaptiveness of personalism [*suki* and allied relationships—*Author*] for development (such as red tape, erosion of bureaucratic honesty, encouragement of corruption, undermining of public morality), personalism may on balance provide functions without which development could not take place (without a rather complete structural and ideological revolution). The advantages of Filipino economic personalism constitute another challenge to the ethnocentric fallacy that only Western social forms and psychological conditions give rise to incipient and more advanced economic development.'

The Fiesta. Social attitudes in the Philippines also have their non-materialistic aspects; dominant in this respect is the *fiesta*. In more sophisticated and wealthier urban communities the *fiesta* may be considered part, albeit an important part, of a social round that has much in common with Western society. But, to the masses, and especially the rural inhabitants, the *fiesta* represents the most important by far of the few opportunities they have to relax from the daily routine. Other opportunities, however, are not lacking: they focus upon weddings or the baptism of a child of a well-to-do member of the rural community. On such occasions, the entire village joins the feast. Even wakes and the nights of vigil over the dead are causes for some entertainment. With these may be included the feasts celebrating the confirmation of a child, the death anniversary of a beloved grandparent or the homecoming of a relative just back from a long sojourn in the United States.

There is no day in the year in the Philippines that a city, a town or a *barrio* does not turn out to celebrate the feast day of a patron saint. Though its religious inspiration was derived from Spain, the *fiesta* has roots in early Philippine societies. Among the non-Christian Filipinos, from the Ifugaos of the Mountain Province to the Tausugs of Sulu, the occasions that call for feasting are no different from those that inspire a banquet in a Manila district. The difference in celebration is chiefly a difference in mode.

The *fiesta* is both a sacred and secular affair. Its centre is usually the town *plaza* from which a carnival spirit is radiated as far as the neighbouring towns and villages and from which a stream of guests is drawn (the success of a *fiesta* almost always is determined from the viewpoint of numerical participation). Though the *fiesta* may be a week long, the principal events come on the eve and on the feast day of the patron saint. During the eve of the feast day (the *vesperas*), lights go up in the plaza where the paraphernalia of a small carnival has found reservation. The early part of the evening may be taken up by an amateur night contest or children's programme sponsored by the local parent-teacher association; this may be followed by a dance. Here will be proclaimed and crowned the queen of the *fiesta*, usually the daughter of a well-known citizen. The queen is usually crowned by a popular political figure who finds the occasion opportune for an appropriate speech.

The day of the feast begins with early masses and is followed by contests of skills at the *plaza*. There may be wrestling matches, a contest in native fencing or even a *carabao* race. The priest may be a little late at joining his guests at the rectory as the day is usually picked by many parents for christening. By now, toward noontime, guest-calling may turn into guest-grabbing when every stranger in the street is forcibly led to a food-laden table and cajoled to partake of the feast despite his protestations that he has finished eating (as with the *fiesta* as a whole, the success of a banquet rests on the number of participants).

National Attitudes. Unlike many underdeveloped countries, political nationalism has not been a striking force in the history of the Philippine Islands. The 'benevolent colonisation' by the United States nipped in the bud the nationalistic elements which were gaining strength as over three centuries of Spanish colonisation drew to a close. American encouragement of national pride successfully channelled nationalistic fervour and removed the dangers of political fragmentation which the physiographic features of the archipelago encouraged. National pride was consolidated by the universal rejection of Japanese aggression. Today, the symbolic Juan de la Cruz (Mr. Average Citizen) is first a Filipino; only second is he a Bicolano, or a Cebuano, or an Ilocano. Filipinos are proud not only of their independence but also of their unity.

National pride, like the Filipino's individual pride, unfortunately has its bad points (the furore created over the relatively unimportant Sabah issue is a case in point). From a geographical point of view it is the over-stressing of economic independence to the detriment of various sectors of resource development, commerce and industry which is of most consequence. Economic nationalism takes the form of minimising the influence of foreign firms (Filipino Chinese being regarded as foreign in this context) by intensive Filipinisation of the economy. Overt Chinese domination of industry and commerce is now the exception but in fact their indirect control by means of 'dummies' and other devices continues to be strong, the companies involved having an estimated market value of £40 000 000. Truly foreign firms have been less successful: in some sectors, such as the rice and maize industry, further growth in alien ownership is prohibited; in others, such as public utilities, it is restricted by law. Only American companies have successfully penetrated the Philippine economy; twenty-four of the fifty largest corporations are U.S. owned, notably in the mining and lumber industries. Here, as in certain other fields, Americans were granted parity rights with Filipinos. This constitutional provision is gradually lapsing but in respect of certain resource developments, notably involving minerals, the rights have been extended to 1977.

International Attitudes. The American heritage strongly conditions the external relations of the Republic of the Philippines. Close ties with the United States, a fear of Communism and staunch support of the United Nations are the cornerstones. Sympathy for the ambitions of other developing nations is not absent, but pro-Western policy predominates in terms of both international politics and trade. Until recently also, the Philippines tended to minimise the consequence of its locational and racial ties with the rest of South-East Asia. Developments such as the Vietnam war and the Sabah issue have now caused a major reappraisal of attitudes which parallel those towards Japan, in which past emnity is being replaced by economic interdependence. Before long a realisation of the role likely to be played by Australia in South-East Asia will further contribute to a repolarisation of Philippine international attitudes.

The international attitudes of the Republic of the Philippines are characterised by many unique features that have internal consequences, both national and regional. Intimate relations with the United States are paralleled, though largely in cultural matters, by those with Spain. The latter relationship has its roots in the past; current realities govern the Republic's attitude to Communist states. In this sphere, past fears of Communism have become muted, though far from dissipated, for a variety of reasons of which perhaps the most outstanding are the economic necessity of finding new export markets and the internal political advantages gained from a less marked pro-American policy. Nearer home, conflict over smuggling and the sovereignty of Sabah continues to irritate relations with Malaysia and further hinders efforts at regional collaboration between the country's Malay neighbours. Better relations exist with Indonesia for, aside from religion, the two countries have much in common. However, one point of disagreement is the pro-Western stand of the Philippines, which in a wide context colours relations to a greater or lesser extent with many Asian nations. Envy of the Philippines' priveleged position *vis à vis* the United States and irritation with the superior attitudes sometimes adopted by Filipinos are contributory factors. The present slight though perceptible shift of emphasis away from the United States and towards Asia could in the long run benefit the country's overall international standing.

In terms of trade, international links have been even more dominated by the United States thanks to preferential trading agreements. All the country's sugar exports, most of the coconut oil, tobacco and pineapple, and approximately half the base metals, copra and abacá are purchased by the United States. In return, the United States acts as the nation's principal supplier of machinery, dairy products, paper, automobiles, drugs, grains and cereal products, petroleum and steel. In all, the United States in 1970 supplied almost 30% of all Philippine imports and purchased over 40% of her exports.

The dominance of the United States as a trading partner is likely to end in the 1970s. Even without the impending termination of the preferential trading agreements with the United States, it is inevitable that Japan will emerge as the Philippines' chief trading partner. Already Japan is the destination or source of one third of all Philippine trade. Japanese businessmen currently purchase large quantities of Philippine timber products, minerals and other raw materials and supply textiles, transport equipment and electrical appliances. In the 1970s Australia could become a major trading partner also. Currently Australia supplies meat, wheat and flour, but as yet purchases only small quantities of the Republic's products.

References

1. A. Cutshall, *op. cit.,* p.24
2. For a succinct yet penetrating appraisal of Filipino character see O. Corpuz, *The Philippines*, Prentice Hall, Englewood Cliffs, N.J., 1965.
3. J.E. Spencer, *Land and People in the Philippines: Geographic Problems in Rural Economy*, University of California Press, Berkeley, 1952, pp. 33-4.
4. J.N. Anderson, 'Buy and Sell and Economic Personalism: Foundations for Philippine Entrepreneurship', *Asian Survey,* Vol. 9, 1969, p. 662. See also O. Corpuz, *op.cit.*, pp. 89-92. Attention is drawn to the conflicting demands of public service and family (in its widest sense) obligations with the latter gaining the upper hand as the Filipino places greater value on personal relations even though the resultant returning of favours may not be in the public interest.
5. R.E. Huke, *op. cit.,* pp. 146-7
6. Estimates made in the mid-fifties suggest that above half the agricultural land of the Philippines was owned by less than one tenth of all farm operators (T. Tsutomu, 'Land Ownership and Land Reform Problems of the Philippines', *The Developing Economies*, Vol. 2, 1969, p. 60).
7. With capital investment in industry still relatively unattractive, due to competition from imports and the small size of the domestic market, real estate remains the favoured form in which wealth is held. Such a process is self-perpetuating, the social prestige and political power so obtained lead to a continuing desire for real estate which pushes its price steadily upwards thus creating an additional, financial, motive for the acquisition of real estate.
8. This attitude stems from various causes including the paternalistic/feudalistic attitude adopted by the Church and the Spanish respectively; the feeling of inferiority resulting from the 'little brown brothers' approach adopted by the Americans; the absence of a significant middle class; the stultifying impact of elaborate kinship ties and obligations; and, paradoxically, the relatively efficient practice of democracy; for it is considered that once one has voted (and poll turnouts are high by Western standards) the actions of those elected, however outrageous by Western standards, through the need to return pre-election favours and obtain favours for the next one, are sacrosanct (at least until the next election).
9. V.W. Ruttan, 'Tenure and Productivity of Philippine Rice Producing Farms', *Philippine Economic Journal*, Vol. V, 1966, p. 63.
10. R.E. Huke, *op. cit.,* p. 197.
11. J.N. Anderson, *op. cit.,* p. 656.
12. *Ibid*, pp. 611-68.
13. *Ibid*, p. 665.
14. *Ibid*, pp. 666-8.

6
The Regions

An appreciation of the regional differences which exist in the Philippines is of value, not only to the geographer, but also to other disciplines. In the past much well-intentioned effort by, for example, administrators and planners has failed to make an impact because of an inadequate appreciation of this fact. It is not enough to know that Ilocanos make good settlers in the pioneer lands of Mindanao, there must be an appreciation also of conditions in their homeland which have produced outmigration. Thus, a long and severe dry season coupled with a narrow coastal strip, bereft of good port facilities and providing only limited opportunities for irrigation, greatly constrains the activities of a hard working people handicapped by a reluctance to abandon rice as their food staple. In the Central Visayas, on the other hand, the local inhabitants are confronted with a different environment and hence, different problems; these, as elsewhere in the Philippines, may be compounded by local features. Thus, on Siquijor Island the population not only has to cope with thin and dry coralline soils, reliance on maize, and large families but also more unique features, such as severe infestation by *lantana*, a reluctance of the inhabitants to emigrate, the subdivision of the island into numerous political districts and the physical separation from the parent province of Negros Oriental.[1]

The Philippines, by embracing an archipelago, might be expected to best submit to regional analysis on the basis of selected individual islands. Certainly, many such land masses do possess a considerable measure of homogeneity. Thus, for example, Samar, Cebu, Palawan, Mindoro and Marinduque, to name but a few, would warrant individual treatment if a full scale appraisal were required of Philippine regional geography. Conversely, a superficial regional study would require giving

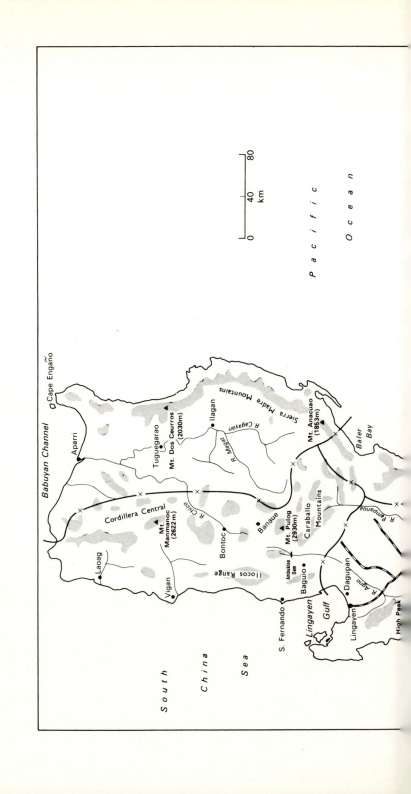

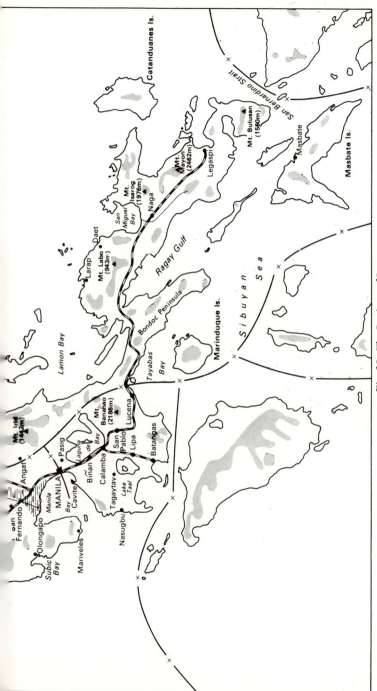

Fig. 30. The Regions of Luzon.

individual consideration only to major islands, such as Mindanao, or major island groups such as the Visayas. The approach adopted in this study is neither full-scale nor superficial. With regional analysis being used as a means to an end, it must balance economy with elucidation. Only a short step down the regional hierarchy is taken: further than the elementary threefold subdivision of Luzon, the Visayas and Mindanao, but not into the ever increasing spatial fragmentation which results from drawing regional boundaries with equal or greater frequency than provincial boundaries.

The varying rural and urban scene of the Philippines therefore is discussed in terms of fourteen regions. Each represents the product of particular people and particular places which are united by ties of geomorphology, religion, government and social attitudes, to name but a few, into the unique geographical entity that is the Republic of the Philippines. The areas so delineated are designed to be of comparable rank in the regional hierarchy of the country and in consequence the homogeneity of geographical features contained within their boundaries is of a similar level.[2] The bases of homogeneity not only may vary from region to region, but also frequently reflect the localised impact of those phenomena discussed in the preceding chapters, and which are considered of fundamental importance to a comprehensive understanding of Philippine geography.

The sub-titles given to each of the following regions not only sum up in a phrase their individual character but also indicate, directly or indirectly, those geographical phenomena of special consequence in the area. Discussion of such phenomena in a regional context is an essential prerequisite to the subsequent chapters in which an attempt is made to assess their contribution to the geography of the Philippines as a whole, and to forecast future patterns of life and livelihood in the archipelago.

The Cagayan Valley—land of unfulfilled promise

The Rio Grande de Cagayan meanders across three provinces—Nueva Viscaya in the south, Isabela in the centre, and Cagayan in the north (Fig. 30). Being essentially of tectonic origin, the Valley's size (eighty kilometres wide and 375 kilometres long) is out of proportion to that of the river by which it is drained, and makes possible a uniformity of climatic and geomorphological features unparalleled in the Philippines. Despite the scope thus offered for large scale development, the Valley is far from attaining its full potential. Natural barriers to development are present, notably flooding, but these are not insuperable. However, the scale of development required is such as to inhibit action on the part of the authorities in Manila and, with politicians from the Cagayan Valley possessing little influence in the capital, the area continues to promise more than it produces, despite recent favourable developments in the production of rice (yields now are second only to the Central Plain of Luzon) and tobacco (higher prices).

The million and a half inhabitants of the Cagayan Valley occupy an area in excess of 25 000 square kilometres. The cultivation on the flood plain and adjacent rolling terrain of various crop associations combining rice (6% of the nation's output volume), maize (11%) and tobacco (29%) is their principal occupation and is supplemented by timber extraction in the surrounding hills. A succession of small service centres is to be found along the banks of the Cagayan River and its principal tributary, the Chico River. Of these, Tuguegarao and Aparri (14 000 persons each in 1970), and Ilagan (10 000 persons) provide the principal focal points of life and livelihood in the area.

The unfulfilled promise of the region is essentially agricultural in nature. The valley has been infilled to depths up to 3250 metres with riverine sediments and the fertility of the easily worked sandy loams in the present floodplain areas is maintained by annual inundations. On the margins of the region substantial quantities of commercial timber, estimated to be in excess of 5 million cubic metres, are to be found. On the other hand mineral resources appear to have little potential, developments of significance being dependent upon the search for oil, which, as yet, has had no success although the geologic history of the region is encouraging.

If the agricultural promise of the area is to be fulfilled the climatic conditions must be mastered. Their homogeneity simplifies matters, but this also means that the twin problems of flood and drought are universally felt. Neither is necessarily a major hindrance: damage caused by flooding such as in 1965 and 1971 is offset by the replenishment of soil fertility while the drought season occurs in the low sun period, and so facilitates harvesting and processing. Nevertheless, considerable advances in agricultural productivity are expected to result from the establishment of flood mitigation and water storage and irrigation schemes.

Mastery of the physical environment alone will not solve the region's problems. The outstanding deficiency of the human environment concerns transportation facilities whose inadequacies are the root cause of much economic and social backwardness. The spine road of the region, Highway 5, is gravelled throughout its length and its river crossings are permanently bridged. But this does not prevent traffic from being dislocated by flooding while poor road maintenance means that it is commonplace for the buses linking the Valley to the Central Plain of Luzon to require a complete overhaul at least once a month. Most other roads are merely dirt tracks passable only in dry weather. The railroad link to Manila was first recommended in 1905. It was not until 1962 that serious work began, only to be stopped soon afterwards for lack of funds. With Japanese assistance the project is hoped to be resumed in the near future but its completion date is still in the realm of conjecture, particularly as an all weather road link would be a more valuable investment. Seaborne traffic is hindered not only by the region's restricted accessible shoreline but also by the fact that the sole port of any consequence, Aparri, is inadequately equipped and located on a coast exposed to severe onshore winds for much of the year. Inland water transport is restricted by the irregular flow of the main streams. Light barges can reach as far as Tuguegarao during periods of high water but in general the streams are utilised to any extent only by small boats. The most efficient method of transport is by air with regular services linking the region's main urban centres with each other and with Manila and the Ilocos coast. The services are well patronised but customers represent only a fraction of those whose life and livelihood would benefit from speedy, efficient and, most important, cheap transportation facilities.

Undeveloped agricultural resources, inadequate transportation facilities and similar deficiencies give the region an air of resignation. Like the underground Didicas volcano off Aparri the region is quiescent and patience and fatalism are ingrained in the Cagayano's character. Talk of the cheap power potential of the Cagayan River, the unexploited fishing grounds off the northern coast, the coming of the railway and the development of the timberlands awake little enthusiasm amongst local inhabitants. Even concrete developments such as the Magat Irrigation Scheme (ultimately to serve 24 000 hectares) have done little to change basic attitudes. Life remains a constant struggle to extract a living from the soil, for only if the rains arrive on schedule will yields be satisfactory, despite the inherent fertility of the soil. Even where irrigation is practised, rice yields are significantly lower than those obtained on the Central Plain of Luzon. Maize thus plays a major role in the diet of the Valley's inhabitants. The region's annual average income—at £125 per family or two-thirds the national average—is similarly well below that which the region ought to sustain.

Despite their hard and often unrewarding life, alleviated only by gambling and the *fiesta*, many inhabitants have known worse times. The bulk of the population are immigrants from the nearby Ilocos coast who have found that in the Valley there is at least land. However, the area's high birth rate is expected to eliminate any significant scope for the pioneer by the mid 1970s. Once that stage is reached, the towns not the countryside will be the focus of migration. But, unless resource development is accelerated in the near future, employment opportunities will be severely restricted.

Like Mindanao, the Cagayan Valley is the land of the pioneer. Unlike Mindanao, however, the terrain is not so much untouched as undeveloped. The interior environment of the Cagayan Valley has probably contributed to the low population densities of the past. But with population pressure increasingly overcoming any prejudice felt by marine orientated Filipinos, the region is steadily filling up. As it does so, rising output and increased political influence should combine to reduce the area's isolation from the rest of the Philippines and attract developmental capital which in turn will further stimulate progress in the Valley. As the area's promise is fulfilled, regional pride will oust gambling, optimism will have replaced fatalism, the roads will be paved and attractive and solidly constructed buildings will have replaced the present day unpainted clapboard structures.

The Ilocos—land of extreme physical and cultural contrasts

The Ilocos region lies adjacent to the Cagayan Valley and occupies the other, western, half of Northern Luzon (Fig. 30). Its interior portion, comprising Abra and Mountain Provinces, extends over some 20 000 square kilometres of highlands and possesses little more than two-thirds of a million inhabitants (commonly known as Igorots or literally, mountain dwellers) of whom a considerable proportion are to be found in self-sufficient but primitive communities. The coastal portion of the Ilocos region consists, in the north, of the provinces of Ilocos Norte and Ilocos Sur, and, in the south, La Union. Only some 9000 square kilometres are involved but they contain a million and a quarter inhabitants, (the true Ilocanos) who are culturally superior to their neighbours inland and possess a better resource base. Diversity is the keynote of this part of the Philippines and here, perhaps more than anywhere else, the geographical pattern vividly reflects the impact of physical extremes and diverse cultures.

Agriculture provides the basis of livelihood throughout the region. It is supplemented in the highlands by tourism, timber extraction and mining, and fishing, salt and textiles in the lowlands. Rice is the principal crop and relies on irrigation for its success; in the highlands the rice terraces are world famous (though covering only an estimated 1% of the region) but the co-operative irrigation schemes of the coastal lowlands most effectively impound and distribute the highly seasonal flow of the streams in whose upper reaches the rice terraces are located. Sweet potatoes rival rice as the staple food of the highlands; maize is grown extensively in the lowlands but is secondary to rice. Much tobacco is also produced on the lowlands, its popularity stemming from a combination of its revenue raising potential (though this is less than in its heyday of a decade ago) and its ability to be grown in the gap between rice harvesting and planting. The chief cash crop of the highlands is vegetables whose production is focused upon the Trinidad Valley, close to Baguio. Here, several thousand hectares are devoted to cabbages, potatoes, string beans, carrots, onions, cauliflower and lettuce.

Fishing in the Ilocos region is centred on Laoag in the north while the salt evaporation beds are concentrated along the beaches to the south and especially in the vicinity of Vigan. A textile industry is centred on a large weaving mill at Narvacan in Ilocos Sur, but much production is derived from cottage industry. Mining is an important and diversified activity of the highlands, and in Ilocos Sur where iron is mined. The mines of the highlands are concentrated in the vicinity of Baguio, and are essentially twentieth-century developments, as has been associated exploitation of forest resources for mining timber. Small scale operators are more numerous where the tourist industry is concerned and for which Baguio and Bontoc are the centres.

With shifting agriculture the mainstay of many highland communities, urban settlement has been restricted and only Baguio (est. 1970 pop. 83 000) can be regarded as truly urban in character.[3] In the lowlands on the other hand a series of small towns are to be found. Of these Laoag, Vigan and San Fernando are preeminent. Most urban centres of the Ilocos coast represent the lowest bridging point of the streams flowing westward from the highlands. As such they command both the flood-plain and the coast line areas upon which economic activity is centred. The lower slopes of the interfluves support some agriculturalists also but as one proceeds inland the terrain increasingly confines the population to the ever narrowing valley floor. It is not until the highland interior is reached that the pattern changes. Here, the virtual absence of alluvial soil makes any area of lesser slope of value whether located at valley bottom or crest. The resulting diffusion of population is accentuated by the presence of a large *kainginero* element.

Physical conditions in the highlands are not conducive to any sizeable density of population. In particular, soils are usually skeletal in nature and even under the most favourable conditions gravelly or sandy material predominates. The lithological characteristics of the parent material, too, restrict soil fertility, a state of affairs which depresses yields on the alluvial areas downstream. The complexity of the geologic structure causes any mineral wealth to be well scattered, while centuries of *kainginero* occupation have decimated and fragmented both the pine forests of the highlands and the monsoon dipterocarp forests of the lowlands. Widespread soil erosion, too, has followed in his wake. The unique climatic characteristics of the highlands provide some compensation, for Baguio has become established as the nation's summer capital and an internationally known resort centre. In contrast, the climate of the Ilocos coast is by no means unique and its exposure to the summer monsoon makes for a marked seasonality of precipitation with all its attendant problems. However, the rapid increase in elevation inland makes annual rainfalls in excess of 250 centimetres commonplace and offers many as yet largely untapped opportunities for hydroelectric power generation.

In the face of the region's physical restraints the Ilocano and Igorot can only survive by cultivating the virtues of frugality and industry, characteristics which are exemplified by the Ifugao rice terraces. Most Igorots represent the remnants of the first wave of Malay immigrants who arrived several thousand years before the birth of Christ, some of whom settled on the Ilocos coast. In succeeding centuries they were gradually forced to the less hospitable interior by more sophisticated immigrants, also of Malay stock but strongly exposed to Hindu culture. To further complicate the cultural background, Spanish influence had a profound impact along the Ilocos coast, but little influence in the highlands where problems of terrain and the bellicosity of its inhabitants were largely insurmountable. The combined effect was to produce two very distinct cultures. Today, however, population pressure on the coast and improved communications have both increased the mobility of the local populace and exposed them to outside influences. In consequence, a reduction in cultural diversity is becoming apparent.

The effect of these developments upon the Igorot is reflected in the steady Christianisation of the population and their thirst for education. The advent of tourism has increasingly meant that the practical value of the more primitive features of the agricultural setting is essentially a commercial one. Nevertheless, shifting cultivation remains a feature of the landscape and, despite the efforts of missionaries, headhunting has not been entirely eliminated. Violence of another sort forms part of the cultural pattern of the lowlanders, the true Ilocanos. Politics not marriage customs are responsible for the bloodletting in this case but such sensational occurrences should not mask the fact that in other fields the Ilocano has made significant progress. Economic development in the area has been stimulated by the success achieved in many walks of life by Ilocano exiles. Population pressure causes many thousands to leave the area annually either on a permanent or temporary basis, and the Ilocano's capacity for hard work is a distinct asset in a tropical country such as the Philippines. With the Ilocos coast unable to feed itself, occupational diversity is essential and the political strength of exiles in Manila and the steady inflow of funds from relatives who have left the area have done much to stimulate economic growth. Contributing to the sophistication of the Ilocano is the above average degree of urbanisation in the area, which is in marked contrast to that of the highlands. Towns too are rarely just agricultural service centres and the intermixing of agricultural, industrial and fishing elements does much to promote social and economic progress.

In common with most parts of the Philippines, much remains to be done in the region. Food output is capable of significant increases, for irrigation schemes, although widespread, are far from universal in suitable areas, and even where in operation are rarely suitable for double cropping of rice. The political pull of the Ilocano is well illustrated by the above average transportation system, but it too is far from perfect; for example, an extension northward of the railway which terminates in La Union and improved road standards would greatly increase efficiency.

Compared to the adjacent Cagayan Valley, however, the Ilocos region has an air of dynamism. This is perhaps most applicable to the coastal area, but, as has been demonstrated, change is an increasingly important feature of life and livelihood in the highlands. Furthermore, the highland inhabitant is not inhibited, as is the case with the Ilocano, by traditional attitudes to, for example, religion and politics. In economic terms, the Ilocano has a bright future but in less material matters the Igorot could prove his superior. Thus, the diversity of the region's geography is likely to be perpetuated though in an ever changing and ever converging manner. In the past the contrasts within the area resulted in disunity; for the future, complimentarity will be the keynote as Igorot and Ilocano benefit from each other's progress.

Central Luzon—the economic heartland

The five provinces of the Central Plain of Luzon—Pangasinan, Nueva Ecija, Tarlac, Pampanga and Bulacan—together with the hill country of the provinces of Zambales and Bataan to the west, an area in excess of 20 000 sq. km, are the home of more than five million Filipinos. Of these some 90% are to be found on the Central Plain itself. The hill country has few attractions for settlement and is perhaps best known for the battlefields of the Bataan Peninsula and of the neighbouring island of Corregidor (Fig. 30).

The Central Plain normally produces nearly one quarter of the nation's rice, and, more important, supplies 40% of the rice required by the food deficit areas of the country. In addition, one fifth of the nation's sugar exports originate here as well as lesser quantities of other agricultural products such as groundnuts, tobacco and vegetables. The largest area of fishpond culture in the Philippines, too, is to be found stretching through the provinces of Bulacan and Pampanga. Another natural resource exploited in the area of the Central Plain is the deposits of the limestone, shale and silica of Bulacan which support two large cement plants. The region's resource basis is augmented by the mineral wealth of the hill country which is centred on the world's largest refractory-grade chromite mine at Masinloc. Salt, fish and rice are the other mainstays of economic activity in this part of the region.

The hill country possesses few towns but they are numerous on the Central Plain. Tarlac, with about 50 000 persons, is dominant, its central location being a distinct advantage, raising it, despite the pull of nearby Manila, to near inter-regional rather than provincial status. Of the urban centres in the latter category, San Fernando (Pampanga) and Dagupan (Pangasinan) are preeminent with estimated 1970 urban populations of 80 000 and 70 000 persons respectively. Even larger urban centres, Angeles (Pampanga) with an estimated 134 000 persons within the city boundaries in 1970 and Olongapo (Zambales) with 100 000 persons, owe their development to nearby military bases rather than to any agricultural service functions which are the mainstay of most of the region's settlements.

The Central Plain is essentially a sea of alluvium through which the occasional volcanic cone protrudes. The largest of these, Mt. Arayat, commands the southern portion of the Plain but, like the belt of undulating terrain which occurs in the south-east corner, the volcanic cones offer little obstacle to movement. In contrast to the west, an abrupt break of slope, similar to that which defines the region's northern and eastern boundaries, clearly indicates the presence of the hill country. Thanks to north-south alignments of the deeply ravined Zambales Mountains, it provides a very effective barrier to the south-west monsoon. Thus, whereas the western portion of the hill country receives more than 250 centimetres of rain per annum, the comparable figure for the Central Plain is less than 200 centimetres. The Central Plain, however, is fortunate in being drained by a succession of streams debouching from the highlands to the north and east; the Pampanga River is the most important of these. Their length and profile contrast markedly with the streams of the hill country: the latter are short and almost entirely youthful in character; the former have short headwater stretches, long, meandering middle courses, and enter the sea through extensive swamplands. It is only in the swamplands that the indigenous vegetation of the Central Plain is preserved to any extent. In the hill country, however, although *cogon* grasslands are widespread, dense stands of dipterocarp forests still clothe the upper slopes which have not yet attracted the *kaingineros*.

Despite the favourable physical environment, economic activity in the region has its problems. Dominating all others are those related to agriculture which are exemplified by the activities of the Hukbalahap (Huks) centred on Pampanga Province. With tenant farming prevalent throughout the Central Plain, associated indebtedness and fragmentation of holdings as a result of population pressures, coupled with the seasonal nature of employment on the sugarlands, have meant that the agricultural workers remain faced with a bleak future, since government sponsored efforts to find a solution have made little headway. Originally Communist directed, but supported nonetheless by the peasantry who put economic advancement before their Catholic devotion, Huk activity now maintains its impetus more through its ability to demonstrate to the masses that it can achieve material gains rather than political recognition. Thus, it is claimed, Huk supporters are protected from unscrupulous landowners, mercenary Chinese middlemen, *carabao* rustlers and corrupt government officials, and that the most effective way for an area to secure government aid for irrigation systems, schools, bridges, roads, etc., is to become a centre of Huk activity. There is no doubt an element of truth in these claims and it is unquestionably true that agrarian unrest will continue in one form or another until much more effective land reform and wider employment opportunites materialise.

The juxtaposition of material abundance and social impoverishment is by no means confined to Central Luzon, but it is perhaps most evident here. Despite having one in eight of the work force unemployed (and even greater numbers underemployed), the region can produce abundant evidence to support its claim to be the economic heartland of the nation. Both in terms of motor vehicle densities and per capita consumption of electricity, the region is second only to metropolitan Manila, while it easily leads in the utilisation of HYV rice seed. Infrastructure facilities are well developed by Philippine standards; there are some 500 kilometres of rail track linking all the larger settlements and the Central Plain is also traversed by two all weather north-south highways.

Expansion of irrigation facilities is continuous—irrigation is essential for double cropping of rice and, if applied properly, results in a fourfold increase in yield per hectare. Recent developments have been dominated by the Angat hydro-electric scheme which, in addition to providing irrigation water for over 100 000 hectares, has doubled local power capacity and provided sufficient drinking water to meet Manila's requirements for the next decade. Further expansion in the 1970s has been assured by a World Bank loan equivalent to over £14 million for the construction of facilities to serve about 80 000 hectares. Developments include a dam on the Pampanga River, rehabilitation of existing irrigation systems serving 48 000 hectares and the construction of new systems to serve a further 32 000 hectares.

Most parts of the region can boast that family incomes are above the national average while in addition two provinces—Bulacan and Zambales—rank amongst the most literate in the country. Unfortunately, the average family rarely exists; the middle class is largely squeezed out between the mass of the impoverished peasantry and small number of wealthy landowners, merchants and politicians. The common *tao*, whether he be a Tagalog or Pampangueño from the south and centre or a Pangasinenso or Ilocano from the north, often may have but a limited cultural and social horizon notwithstanding the greater technical sophistication engendered by the introduction of HYV rice. Apart from family and *fiesta*, life can be a continual struggle; to him 'every meal is a fulfilment and to go to sleep on a full stomach is a triumph and every windfall is a miracle.'[4] With only very slow progress made by the Land Bank in releasing the £40 million called for in the Land Reform Code, such extremes of poverty will continue to be a feature of the economic heartland of the nation.

Manila—the core of the nation

Manila is located at the southern tip of the Central Plain of Luzon (Fig. 30) but its socio-economic importance is such that it forms a distinct regional entity. Metropolitan Manila is the home of some 3 million persons. Their activities, whether they be industrial, commercial, financial, cultural, educational, governmental or religious, provide goods and services for the entire nation. Manila also provides the only link of major significance with the outside world.

The metropolis came into existence as a port, for the resource base of its immediate hinterland consisted of little more than a protected anchorage at the mouth of the Pasig River, a stream which emptied into the landlocked, hence sheltered and easily defended, Manila Bay. For its growth Manila has owed more to situation than site. The narrow lowland between the Bay and Eastern Seaboard Range provided a land corridor between the Central Plain and the volcanic lowlands of South-Western Luzon, two of the most important agricultural regions not only of Luzon but of the whole archipelago. To the trade potential thus tapped was added that of the islands' principal external link, with South China, which Manila was well positioned to exploit. The port was already flourishing prior to the arrival of the Spanish but it was they who ensured its dominant rule; internally, by making it the centre of the colonial administration; and externally, by designating it the home port for the Manila Galleons, which constituted the colonists' sole link with the New World, and Spain itself. The American administration brought new vigour and material improvement to the city's stagnating port facilities while living conditions reaped the benefit of intensive activity in the fields of sanitation, electricity and other public services. Much of this enlightened effort was a casualty of the Japanese invasion, for by the end of the war three-quarters of the city had been razed to the ground. In the post-war period, the ever swelling population of the metropolis has continually limited the effectiveness of reconstruction work and subsequent new development, but this should not obscure progress achieved.

Today, metropolitan Manila spreads far and wide from the original two native towns—Tondo and Maynilad—located on either side of the mouth of the Pasig. The former site is now occupied by slums and the business quarter, the latter by the nucleus of the Spanish city, Intramuros. Intramuros itself forms one of fourteen districts which made up the original City of Manila. Today, seven major suburban communities have to be included to encompass the official Greater Manila metropolitan area (see Fig. 4), and a further eight smaller communities should also be taken into account when discussing the region.

The port still dominates the life of the City itself. Furthermore, with a captive hinterland embracing the whole of Luzon, Mindoro and Palawan, Manila is the country's leading inter-regional and regional trade centre as well as being by far and away its most important international outlet.

Manila is the principal market for regional surpluses and the major source of locally manufactured consumer goods. Intimately connected with the port is the central business district which encompasses the districts of Binondo, Quiapo and San Nicolas and has now spread south of the river into the districts of Ermita and Malate (Fig. 4). The districts fringing this core area of the City are becoming less suburban and more commercial in character. This process has reached an advanced stage in Tondo, the largest and most densely populated district. Here, the railway terminus and the inter-island port terminal attract a variety of small scale commercial and industrial establishments and also the major produce and clothing markets, which profit both from the transport facilities and from the abundant supply of cheap labour drawn to the area by the low rentals of this poorly drained location. Paco, San Miguel and Pandacan districts are in complete contrast to Tondo, these old established residential areas having the Presidential Palace of Malacañang as their original focal point. Sampaloc and Santa Ana districts also fall into this category.

The communities peripheral to the City of Manila are symptomatic of the post-war growth of the metropolis. As such, their very structure differs from that of the City districts with the shopping centre increasingly rivalling the church and the *plaza* as the focal point of the community. Each, too, is an independent political and administrative unit. The largest, Quezon City, has been designated as the capital of the Philippines and, although few government bodies have yet moved there, ancillary institutions, notably the University of the Philippines, are well established. So, too, are several large, privately financed housing estates and associated shopping complexes which are becoming an increasingly important feature of the suburban scene in general.

Unlike the City of Manila, most of the suburbs have a strong industrial base. The place of employment for many suburban dwellers, and especially white collar workers, lies in the City, but ever increasing numbers are employed in the industrial establishments which have sprung up on the periphery of established settlement. Here, cheap land is the attraction, especially if located adjacent to major highways. Industry is not the only new activity attracted to such locations; market gardening, poultry and egg production, and allied agricultural activities designed to satisfy the demands of the ever growing metropolitan population form an important component of the land use pattern. The suburban scene is, therefore, complex, comprising:

'concentrated poblaciones, scattered barrio nuclei, and open agricultural lands, so large industrial sites have lain chiefly in the open tracts between the older settlement sites. Light manufacturing and service industries have accumulated in and on the fringes of the old poblaciones, along primary roads, and in the various old barrio centres. There is relatively little functional concentration in the development of industrial districts; instead there is a tendency for all kinds of industries to scatter into all suburbs according to criteria related to local transport, the availability of land, and various other temporary determinants of site location ... The siting of new residential tracts, communities, and zones in the suburbs is somewhat competitive with the siting of industrial activities, as large open-land areas are needed by both. The result is a rather mixed residential, commercial, industrial urban complex in which all kinds of activities occur in all kinds of areas.'[5]

TABLE 23

Manufacturing Employment in Greater Manila, 1960 (by type of product and political unit)

Major Product Group	Manila	Quezon City	Caloocan	Manda-luyong	Malabon	Makati	San Juan	Parañaque	Pasay	Navotas	Total Employment
Food and Beverages	8810	2237	1711	932	1445	451	587	963	306	252	17 694
Tobacco Products	2790	1048	484	566	1098	94	—	1643	1449	—	9172
Textiles and Related Products	5549	5848	1501	3276	1951	796	1045	53	526	107	31 232
Lumber	6526	629	322	196	30	221	118	—	110	89	8241
Paper	10 428	2401	481	374	240	63	281	—	106	81	14 455
Rubber and Leather	976	221	1554	28	758	2256	—	—	50	80	5923
Chemical and Allied Products	3822	962	458	1776	330	967	189	18	24	—	8546
Non-Metallic Products	1666	221	525	1101	329	242	650	10	4	31	4779
Metal Products	4170	2230	2515	1585	437	271	98	117	271	162	11 856
Machinery	5242	408	591	327	149	1531	1000	—	—	—	9248
Transport Equipment	3532	310	377	356	—	110	63	3	16	103	4870
Miscellaneous	1132	444	687	56	439	45	10	502	360	13	3688
Total	65 223	16 959	11 206	10 573	7206	7047	4041	3309	3222	918	129 704

Source: Central Bank of the Philippines. Based on 1960 Census data; that for 1970 not yet published.

Industrial activity has long been a feature of the metropolis[6] but City based enterprises were usually small scale operations, their size being restricted in the pre-war period by limited land and capital, and a limited market both in terms of size and spending power. Emphasis, therefore, was, and still is, upon basic products such as food and beverages and textiles: in 1960 the latter category accounted for 24·1% of the manufacturing

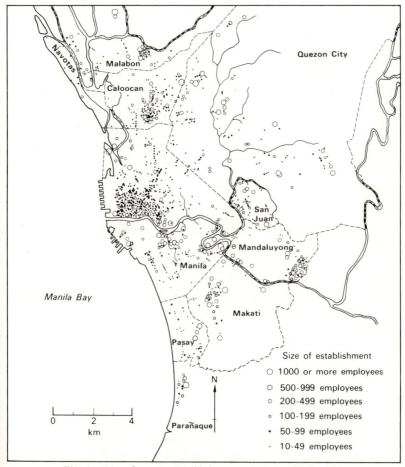

Fig. 31. Manufacturing Establishments, Greater Manila, 1960

work force in Manila while the former accounted for a further 13·7% (Table 23). In the post-war years both the size and sophistication of manufacturing have increased in parallel with the economic advancement of the country and the government policy of import substitution. Heavy industry, including steel fabrication, chemicals, motor vehicle assembly and food processing, has been attracted to waterside locations. Notable in this respect have been the banks of the Pasig River in Mandaluyong upstream from the traditional centre of river orientated industry at Paco. Today, little riverside land remains available for development, but this has coincided with the supplanting of the river barge by road transport as a link with the port areas. The emergence of light industry based on imported raw materials and located along the principal highways is especially evident in Malabon, Caloocan, Mandaluyong and Makati (Fig. 31). In the two last mentioned areas substantial industrial estates have grown up and it is here that most foreign manufacturing companies have their factories. The nature of light industry in the suburbs is diverse, but outstanding are textiles, furniture, rubber products, electrical appliances, food and beverages and pharmaceuticals.

Although industrial development holds the key to much of the growing importance of the metropolis (Greater Manila accounts for one half of the country's manufacturing work force), one must not lose sight of the fact that its other functions have not diminished in significance. Filipinos look to Manila for leadership, knowledge and spiritual guidance as much as material benefits in the form of motor vehicles, medicines and clothing. Its dominance is absolute and, like Bangkok and Saigon, its counterparts across the South China Sea, it is an outstanding example of a primate city, the core area of a nation.

South-Western Luzon—volcanic creation and metropolitan annexe

In South-Western Luzon there intersect the two principal mountain building axes of the country resulting in a dominance of vulcanism, past and present. It is here more than in any other part of the archipelago that the consequences of living in an area of crustal instability are evident. Active vulcanism and strong seismic activity periodically have disturbed the pattern of life and livelihood in South-Western Luzon but they have not prevented its inhabitants from making good use of those beneficial physical features created, nor from exploiting the locational advantages of the region. Essentially, South-Western Luzon comprises the three provinces of Batangas, Cavite and Laguna (Fig. 30). If small adjacent portions of Rizal and Quezon Provinces are included on account of their geographical affinity, the region covers some 7000 square kilometres and possesses a population in excess of 2 million persons.

Proximity to Manila has had a profound effect. The agricultural base, for example, exhibits a degree of diversification unparalled elsewhere in the country. The region's versatility in producing substantial quantities of rice, coconuts, sugar, fruit and maize is, however, stimulated only in part by the demands of the metropolis; the fertile soil and varying terrain created by vulcanism, and by the transitional nature of the climate, are equally potent factors. Perhaps more important is the industrial and commercial diversity of the region. In the north, the economic development of Greater Manila has caused an overflow of light industry and associated activities into the region. Other parts have benefited also from the growth of Manila. The establishment of petroleum refining in Batangas Province exemplifies the ability of the region to provide suitable sites for activities orientated towards the metropolis but not necessarily dependent upon its intimate proximity. In like manner, a variety of smaller industrial enterprises such as shoes, clothing and salt evaporation have become of importance in the region's trade centres in recent years. Finally, mention must be made of the tourist potential of the region for which again proximity to the metropolis is advantageous. A marked increase can be expected in the already sizeable number of visitors to the area both originating in Manila and overseas, anxious to view not only such spectacular physical features as Lake Taal but also points of historical interest such as the ancestral home of Jose Rizal at Calamba (Laguna Province).

In physical terms, the landscape of South-Western Luzon is one of marked contrast between the alluvial lowlands and the volcanic uplands. This is a state of affairs to be found in many parts of the Philippines, but it has few rivals in terms of scale.

Only along the west coast of Batangas in the vicinity of Nasugbu, where old sedimentary rocks outcrop, is any material of non-volcanic origin exposed. At the opposite, eastern, side of the region, the massive pile of Mt. Banahao (2188 metres) dominates the skyline. Lesser volcanic uplands occupy the southern portion, especially to the east of Batangas town. The northern part of the region is dominated by the expanse of Laguna de Bay into whose northern reaches poke two spurs of the Eastern Seaboard Range. Lake Taal, roughly one third the area of Laguna de Bay, is the principal physical feature of the centre of South-Western Luzon, and here the intermixing of lowland and highland becomes most complex. In most parts of the region, upland areas are connected by often lightly consolidated deposits of ashes, cinders and tuffs fanning out from the originating volcanoes. On the upper, steeper reaches of these deposits drainage channels may be markedly incised, for the region, and especially its eastern half, experiences heavy rainfall for all but the first few months of the year. The shortness of the dry season, coupled with the favourable characteristics of the volcanic soils (light, deep and well drained) are particularly encouraging to the establishment of the coconut, that ubiquitous feature of Central and Southern Philippines.

The appearance of the coconut as an integral element of the rural scene should not disguise the fact that rice is the dominant farm product:

'rice and fruits other than bananas are grown on about two-thirds of all farms; bananas show up on about half of the farms; coconut is also grown on about one farm in two; and corn (= maize—*Author*) is produced on about one farm in four. About half the farms can be classified as rice farms, about 15% as coconut farms, and about 10% as farms engaged in mixed farming.'[7]

Farms are mostly small in size but, thanks to a high incidence of irrigation (albeit related to small scale schemes dating from the late Spanish era) and owner operatorship, productivity is above average and the proximity of Manila encourages a commercial rather than a subsistence attitude towards agriculture. Thus, to cater for local and metropolitan demand the farmers of South-Western Luzon have successfully turned their attention to livestock production, coffee, kapok and flowers.

Apart from agriculture, the fishing grounds of Manila Bay and Laguna de Bay offer a means of livelihood for many inhabitants of Cavite and Laguna Provinces. Small numbers are employed in tourism also but manufacturing and commerce employ the bulk of the non-agricultural workers. Old established handicraft industries, notably the production of footwear and clothing, persist but are now overshadowed by more modern production facilities. These are concentrated in settlements located immediately to the south of Manila, such as Cavite, Las Piñas and Biñan. Here, too, furniture manufacture, salt, fishpond farming and vehicle body works increase the diversity of employment opportunities. Further afield, the other urban centres of the region notably San Pablo (Laguna Province) and Lipa and Batangas (Batangas Province) are essentially administrative and agricultural service centres. As such, their function is somewhat restricted by the proximity of Manila but, on the other hand, industrial development has been stimulated in recent years. This is most noticeable in Batangas where petroleum refining is concentrated but agricultural processing, notably rice and sugar milling, and textiles are the most common sources of industrial employment. Settlements can be of considerable size, that of Lucena being some 70 000 persons in 1970 while San Pablo and Lipa, for example, each are estimated to contain 42 500 and 30 000 persons respectively in 1970.

The geography of South-Western Luzon, like that of the Central Plain to the north, is strongly influenced by the proximity of Manila. In the past this has resulted in intensive colonisation by the Spanish: their influence remains visible today, for example in the irrigation systems and dense road network. Today, a new force emanates from Manila: the northernmost towns of the region are becoming dormitory settlements while in rural areas intensification and diversification of agriculture proceeds in response to the demands of metropolitan markets. These trends appear likely to persist and, with increased industrialisation and tourist activity probable, both the life and livelihood of the region's Tagalog speaking inhabitants can be expected to become steadily more varied and complex. Yet, whatever man attempts to fashion out of the region with the aid of metropolitan stimuli it will be at the mercy of the forces of nature, and of vulcanism in particular.

Bondoc and Bicol—two typhoon lashed peninsulas

That part of the island of Luzon not yet discussed consists of some 28 500 square kilometres in the form of two peninsulas and numerous islands which straggle south-eastwards from the main body of the landmass in the direction of Samar (Fig. 30). The province of Quezon straddles the junction of the region with the main body of Luzon and extends eastwards to encompass the Bondoc peninsula (and also northwards for 250 kilometres to merge with the eastern portion of the Cagayan Valley) in a narrow (average width 30 kilometres), extremely rugged, and densely forested strip. The large Bicol Peninsula which succeeds it is subdivided into four provinces—Camarines Norte, Camarines Sur, Albay and Sorosogon. Of the region's numerous offshore islands, three—Cantanduanes, Marinduque and Masbate—are important enough to form separate provinces. The Bicol Peninsula is the dominant component of the area not only on account of its size and economic importance but also by virtue of its comparative areal compactness and cultural and linguistic coherence. Of the region's total population of 4·2 million persons estimated for 1970, 62% live in the Bicol Peninsula proper. Like South-Western Luzon, vulcanism has a fundamental role to play in the geography of the Bicol Peninsula, but it is another equally violent physical phenomenon—the typhoon—which has the greatest impact upon life and livelihood not only in the Bicol Peninsula but also throughout the region.

Typhoons are very much a commonplace occurrence. The region experiences 40% of the storms carrying high velocity winds in the Philippines with the result that agriculture, the cornerstone of the economy, continually suffers substantial reductions in the output of its mainstays—coconut, abacá, rice and maize. The result is a standard of living for the region's inhabitants which is not commensurate with its resource base. This is characterised by sizeable areas of fertile soil, extensive forest areas, valuable mineral deposits and important fishing grounds. Typhoons, of course, are only one of the retarding factors; there are, for example, also the problems posed by the physical barriers to communications, inadequate infrastructure and the character of the people.

The physical features of the region are many and varied. In

Quezon Province the rugged Sierra Madre leaves little space for the development of alluvium even in the lower lying Bondoc Peninsula but, being exposed to the moist north-east trade winds, its slopes support extensive stands of timber. On the Bicol Peninsula similar drought-free climatic conditions, which can result in an annual precipitation in excess of 500 centimetres, coupled with more fertile volcanic soil, have created an even more luxuriant arboreal cover. However, in this part of the region infilling between volcanic peaks and former islands has created a series of fertile lowlands which formed the base for subsequent felling of timber in the more accessible areas. Today, some two-thirds of the peninsula is classed as arable and in volcanic areas the forests are confined to the upper slopes, except where recent volcanic activity such as that associated with Mt. Mayon has removed most of the vegetation cover. The forest cover and associated *cogon* grassland is more in evidence in the less fertile non-volcanic areas where hilly terrain, based on uplifted sedimentaries, exists. Such conditions are most evident on the mainland in Camarines Norte but similar vegetation predominates on Catanduanes Island, though here the hilly terrain has developed on basement-complex materials, mainly serpentines and schists. Masbate Island can be regarded as a continuation of the Bondoc Peninsula. It is better favoured in respect of gentle slopes especially in its southern half. This means that much of its forest cover has been removed and replaced by *cogon* grassland which supports cattle raising of some significance. The mineral wealth of the region is concentrated in Camarines Norte, where high grade iron ore is mined at Larap, and on Marinduque, where copper mining has become preeminent in recent years. The non-volcanic rocks of the region also contain variable deposits of coal, pyrites, gold and copper. On a national basis the mineral output of the region is second only to that of Mountain Province. Throughout the region coastal dwellers have easy access to often rich fishing grounds, and significant commercial production occurs in various localities, accounting for about one tenth of the total for the Philippines.

The climatic conditions which have favoured forest growth and the vulcanism which has contributed much to soil fertility also have their drawbacks. The effect of typhoons on agriculture has been mentioned already, but such storms also can make sea transport difficult for much of the year, especially along Pacific shores where exposure to the north-east trades forms an additional hazard. Such disruption of communications is a serious matter, for the configuration of the region hinders other forms of transport. The Bicol Peninsula has both rail and road links to the main body of Luzon but their effectiveness is as tennous as the thirteen km wide Tayabas Isthmus that joins the region to Luzon proper. Within the region, transport is but one of the items of infrastructure which suffers from a lack of capital. Thus, not only are port facilities often primitive in the extreme and the road network poorly developed outside the Bicol Peninsula, but the absence of any dominant administrative and commercial centre leads to a paucity of social and economic institutions such as tertiary education and banking facilities. There are only two urban centres on the mainland of any size—Naga and Legaspi. The former (estimated 1970 population 79 000 persons) can be regarded as the commercial capital of the region, the latter (estimated 1970 population 75 000 persons) serves as its principal port. Masbate, the capital and commercial centre of Masbate Island, also contained more than 15 000 persons in 1970. All other urban centres in the region function predominately as agricultural service centres, fishing ports, notably Daet and Sorosogon. The Bicol Peninsula was once noted for shipbuilding, industrial development is now of little significance, for the purchasing power of its inhabitants is limited and access to other markets is restricted by poor communications. Abacá provides an exception to this state of affairs as, stimulated by government assistance, the local handicrafts industry remains active while in complete contrast, both Philippine and Japanese financiers have expressed interest in creating an abacá based paper pulp making industry (one such plant already has been approved by the government for establishment in Camarines Sur).

Despite the often violent forces of nature which surround him the Bicolano is, by Filipino standards, characterised by a passive outlook on life. Generally, the soil and sea can provide for his immediate needs, and can offset any deficits occurring elsewhere in the region. The destruction wrought by typhoon and volcano is thus accepted fatalistically. His character has also been moulded by the relative isolation of the region; although Tagalog influences are strong in Quezon Province, a distinctive

Bicol language and associated cultural pattern coincides closely with the boundaries of the remainder of the region.

The inward looking, subsistence mode of life characteristic of most inhabitants can be expected to change but slowly. Few major developments appear imminent to improve the region's infrastructure and, with efforts by private enterprise to develop commercially the resource base being scattered and usually on a small scale, only a gradual improvement in living standards appears likely. The life and livelihood of the region will thus continue to be governed by the crops, the fishing and the typhoons.

The Eastern Visayas—the neglected islands

The Portuguese circumnavigator, Ferdinand Magellan, made his first Asian landfall in this part of the Philippines, but the islands of Samar and Leyte today are no longer in the forefront of history. A combination of location, restricted natural resources and limited political influence has placed this region outside the mainstream of Philippine commercial and cultural life. The result is a pattern of agricultural occupance traditionally orientated toward subsistence which is unlikely to alter radically in the near future, the Pan-Philippine Highway and Tacloban iron mining developments notwithstanding.

The unsophisticated way of life in the Eastern Visayas makes for an absence of urban settlement and associated infrastructure. An estimated 2·8 million persons dwell upon the region's 21 000 square kilometres but only in the Leyte Valley (Fig. 32) is there any marked concentration of population. Here, about one fifth of the region's population is to be found, their presence largely contributing to the size of nearby Tacloban (estimated 1970 urban population 50 000 persons). In the past, the physical closeness of Samar and Leyte (only 600 metres apart in places) and the sharing of a common dialect (Waray-Waray) have led to a uniformity of culture patterns. Increasingly, however, the western portion of Leyte has become orientated towards Cebu thanks to an influx of migrants across the Camotes Sea and the subsequent strengthening of commercial and cultural links with what is the second city of the country.

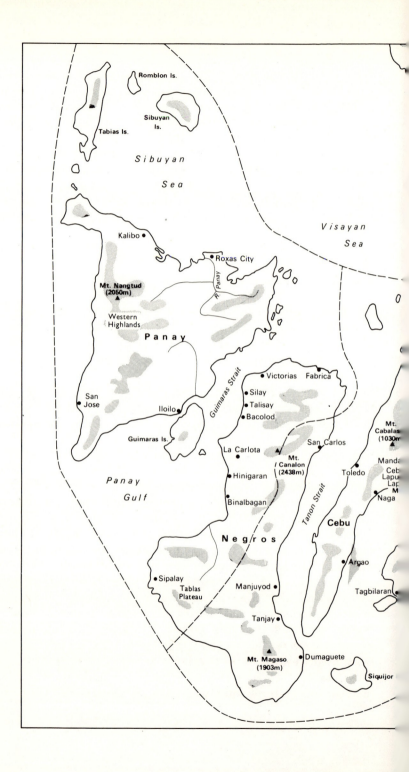

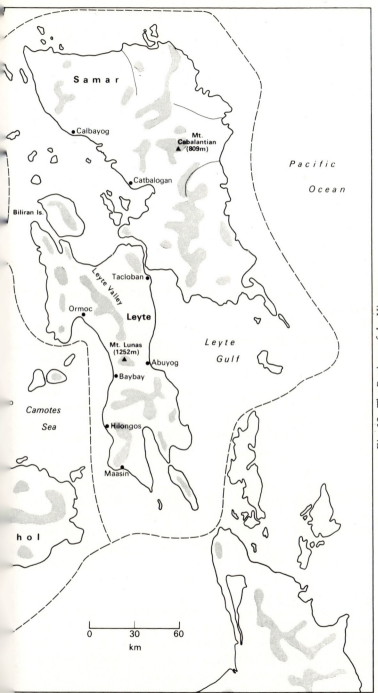

Fig. 32. The Regions of the Visayas.

Samar is the third largest island of the Philippine archipelago and is formed from a broad, maturely dissected, plateau surface with local relief in the order of 200 to 300 metres. With its short but numerous streams flowing in sharply incised valleys, the island's topography is characterised by low but extremely rugged hill lands. In contrast, the spine of Leyte consists of three mountain systems with maximum elevations in excess of 1250 metres whose rugged terrain is very different in nature to the adjacent Leyte and Ormoc Valleys which have been created out of recent alluvial infilling. Apart from these two vicinities no lowland areas of major significance are to be found in the Eastern Visayas.

The general hostility of the region's terrain is matched by its climate. This is especially so in the eastern half of the region which is fully exposed to the moisture-laden winds which blow from the north-east and the typhoons that advance upon it from the south-east. Even in partially sheltered areas to the west no pronounced drought period exists; annual precipitation levels below 250 centimetres are exceptional while levels above 400 centimetres are not uncommon. Such a rainfall regime, coupled with the constantly high temperatures, has resulted in a luxuriant vegetation cover which, although reduced in extent and quality by man's agricultural activities, still contains large stretches of merchantable timber as yet little depleted by existing logging activities.

Spanish influence and subsequent economic development of the region was handicapped until well into the nineteenth century by the depredations of Moros from Mindanao. With law and order established, population growth accelerated for a while but this did not persist into the twentieth century for the opening up of the American markets to Philippine raw materials brought little benefit to the region. As far as mineral wealth was concerned both Samar and Leyte were poorly endowed, while in terms of agricultural products, soil and climatic conditions proved discouraging. Despite the abundant precipitation, double cropping of rice is exceptional although at least two crops of maize are possible. The lack of suitable lowlands results in the production of upland rice accounting for one quarter of total output; maize is of special importance in Western Leyte, reflecting its links with Cebu. Supplementing or, in time of crop failure, augmenting the two staple cereals are a variety of root crops, notably sweet potatoes, taro and yams; bananas also figure prominently in the local diet. Commercial crops are dominated by abacá and the region produces almost one fifth of the nation's output. The coconut is another widely grown source of farm income but the other important crop, sugar, comes almost exclusively from the Ormoc Valley which possesses the region's only *central*.

Throughout the region the embayed nature of the island's coasts not only encourages marine transportation but causes fishing to rival agriculture as the principal occupation of the inhabitants. Weather conditions often disrupt operations but have not deterred the establishment of commercial enterprises, principally in Western Samar, to supplement the semi-subsistence activities of the inhabitants of the region's numerous fishing villages. Commercial fishing activities in the region, like its other business enterprises, operate on a comparatively small and unsophisticated scale. More impressive operations would be out of context, for local markets are of limited extent and access to those in other parts of the country is not easy. Regular links to neighbouring islands, and Manila, are provided by a variety of air and shipping lines. However, both the means of transport and port facilities leave much to be desired, as does the weather at times. Only Tacloban can offer adequate facilities and as a result it has developed as a local transshipment centre. Its influence would be greater but for the handicaps posed to inland transport by the poor road system, whose inadequacies are felt throughout the region and especially in Samar. By 1973, however, current construction of the San Juanico Bridge (part of the Pan-Philippine Highway) will be complete, thus providing a road link between Samar and Leyte.

The state of the road network (inadequate ballast, poor and often minimal maintenance, impassable in times of heavy rain due to washouts and the collapse of flimsy bridges), highlights one of the main obstacles to economic advancement in the region—the lack of government assistance. Like Cagayan Valley, the Eastern Visayas over the years have had little political influence in Manila. Isolation, a comparatively small population, and a subsistence economy have not helped the region's political representatives who also have had the unhappy knack of all too frequently belonging to the Opposition party. Without political pull and without the natural resources to attract private enterprise, the Eastern Visayas only seem to attract attention when spectacular damage results from a particularly violent typhoon.

For the immediate future life and livelihood in the Eastern Visayas can be expected to continue on its routine and parochial way, at least until the advent of the Pan-Philippine Highway. For the élite of the region—government officials, teachers, fish and produce merchants—air transport has expanded contact with urban life, particularly that of Cebu. But the bulk of the population remains very close to the soil or the sea. The region's urban settlements offer few attractions for apart from Tacloban, only Catbalogan and Calbayog in Samar and Ormoc and Baybay in Leyte had populations in excess of 10 000 persons in 1970 (the estimated 1970 populations of Calbayog and Ormoc are respectively 50 000 and 25 000 persons). Thus, it is not surprising that the region is an area of outmigration—the more ambitious moving to Cebu or Manila and the landless moving south to Mindanao.

The Central Visayas—the pivot of the nation

For most parts of the Philippines transportation problems are an inhibiting factor in the pattern of life and livelihood. In the case of the Central Visayas, however, transportation is an integral and positive element in the character of the region. This state of affairs has arisen from the region's strategic location. Cebu City, the focal point of the region, is centrally situated on Cebu Island which is the central land mass; this bisects the Visayas which, in turn, lie roughly equidistant from the nation's principal land masses—Luzon and Mindanao (Fig. 32). The pivotal attributes of the region have been successfully exploited by its inhabitants who themselves, as a consequence of a unique combination of physical and cultural conditions, possess several outstanding attributes.

In addition to the island of Cebu, the Central Visayas also encompasses the adjacent island of Bohol and the eastern half of Negros Island. The homogeneity of the 15 000 square kilometres involved is derived in part from the essentially uniform geologic and climatic conditions experienced, but principally from the cultural affinities of the estimated (1970)

3.5 million inhabitants. The most obvious expressions of this are a common dialect (Cebuano) and a reliance on maize instead of rice as the staple item of diet. However, Cebuanos also are noted for their individualistic character which, coupled with a well deserved reputation for tenacity and hard work, has enabled them to make above average use of the region's available resources and to exercise considerable influence beyond the confines of the Central Visayas.

Nature has not been liberal in the Central Visayas. Both uplands and lowlands are covered by a mantle of calcareous type sedimentary rocks which are prone to excessive erosion. Natural erosion in the past has not produced extensive lowland areas for, except in Bohol, markedly anticlinal structures form the basis of the islands. The plateau-like structure of Bohol has created significant areas of peripheral lowland along its northern coast; elsewhere karst topography predominates. In recent centuries the ever increasing density of settlement, in Cebu and Bohol especially, has accelerated erosion rates by means of extensive removal of the molave forests formerly found throughout the region. This has exposed the all too common steep slopes to the full effects of a climate which, although characterised by relatively low precipitation levels, still provides 150-200 centimetres per annum, largely in the form of tropical downpours (although compared with the Eastern Visayas, cyclones are infrequent). To add to the problems of soil erosion, the calcareous nature of the soil results in high porosity structures which heighten the effects of the dry period which normally extends from December to May, and which can produce drought conditions between February and April.

Despite their unfavourable characteristics, the soils of the region are inherently fertile; rice production may be handicapped by the unfavourable environment, but its substitute, maize, grows well. Soil 'mining' and other ill-advised agricultural practices cause yields to be below average, but with the possibility of triple cropping in many areas annual production levels are high. If maize is ubiquitous (being found even on 15 degree slopes) rice production is only of significance on small river floodplains and deltas, chiefly in Bohol. Secondary food crops include camotes, cassava, jackfruit, bananas and papayas. Commercial agriculture in the region owes much to the coconut which covers nearly 100 000 hectares, and accounts for 10% of the nation's output, largely in the form of copra. Second only to the coconut is sugar cane; the region possesses four *centrals* (two in Cebu and two in Negros) serving over 10 000 hectares. Fishing operated on a purely commercial basis does occur off the northern coasts of Bohol and Cebu, but it is predominantly a subsistence orientated occupation scattered throughout the region.

The region produces various other commodities—timber, abacá, kapok, etc.,—on a commercial basis, but only those derived from mining are of major significance. Coal mining in Cebu (at Danao and Argao) is currently experiencing a resurgence of activity, the output of its 3500 employees being consumed by a local cement producer. The Toledo (Cebu) copper mines are major employers of labour also; manganese extraction on a small scale occurs on Bohol; and several large solar evaporation salt beds are located at Manjuyod (Negros Oriental).

Industrial activity in the region has its base in maize and sugar milling together with other primary product processing activities such as the copper concentration plant at Toledo and the cement factory at Naga (Cebu). However, in recent years other industrial activities have grown up, notably a coconut oil and copra cake factory, a flour and animal feed mill, a brewery, match and rubber products plants, and vehicle assembly shops. All major industrial enterprises, and the bulk of the smaller ones, are located in and around Cebu City. Many represent branch plant facilities established by Manila based concerns, for the Cebuano *forte* is commerce rather than industry.

Cebu is second only to Manila as a trade centre—its influence extends well beyond its immediate hinterland of the Eastern Visayas and the northern half of Mindanao. Large quantities of manufactured goods are distributed from Cebu, with maize, abacá, sugar, copra etc., as return loads; most air and sea passage routes serving the southern half of the Philippines are orientated via Cebu, which also serves as a port of call for foreign shipping. To further its preeminence in this field a major reclamation scheme is under way which, in addition to providing more space for port and manufacturing facilities, may incorporate a Free Trade Zone.

The locational advantage of Cebu lies primarily in the ease of access it affords to the markets of the southern Philippines as a whole rather than the purchasing power of the inhabitants of Cebu. However, the latter is not insignificant in view of the density of population, and it can be expected to grow in importance as increased industrialisation injects wealth into the local economy. In particular, industrialisation can be expected to add impetus to the population growth of the city of Cebu, although such a development may be retarded by the average Cebuano's aversion to urban life. Thus, apart from Cebu City itself, with a 1970 estimated urban population of 345 000 persons, adjacent settlements such as Lapu Lapu on Mactan Island (60 000 persons) and Mandaue (40 000 persons), and the mining centre of Toledo (25 000 persons), only the provincial centres of Dumaguete (estimated 1970 population 25 000 persons) and Tagbilaran (8500 persons in 1970), located respectively in Negros Oriental and Bohol, can be regarded as truly urban in form and function. In contrast, the average agricultural service centre or mining community in the region rarely contain more than a few thousand persons, the only exceptions of significance being on Negros where the populations of San Carlos and Tanjay exceed the 10 000 mark, largely as a result of the presence of workers in their adjacent sugar *centrals*.

The export of the products of Cebu's industrial enterprises, including coconut and maize oil, flour, soap, ceramics, food and beverages, paper products and glass, serve to supplement the city's long established and thriving entrepôt trade. Typical activities in this sphere include copra buying throughout the southern Philippines and its transshipment for direct export, the grading and baling of abacá from Mindanao and the Eastern Visayas, and the widespread distribution of consumer goods originating in Manila and overseas. Such commercial activities are complemented by a less tangible but none the less very important function which also stems from the pivotal position of the region. This involves *inter alia* the production and despatch of local newspapers and periodicals to meet the demands of some 7 million Cebuanos who now reside outside the boundaries of the Central Visayas. The massive outmigration of Cebuanos has its root in the overpopulation of the region and of the island of Cebu in particular, which is largely a consequence of the limited availability of agricultural land. Migrants have been specially attracted to Mindanao for, in terms of temperament, agricultural inclinations and commercial ties, the Cebuano is well fitted to this pioneering environment. Many other Cebuanos have settled in the Eastern Visayas, in the Manila metropolitan area and overseas, in Hawaii and California.

Passenger traffic thus provides a valuable adjunct to the movement of goods which forms the basis of business for the multiplicity of commercial organisations established in Cebu. The pivotal position of the city automatically ensured its emergence as the focal point of the developing airline services in the central and southern Philippines, and its airport is second only to Manila in volume of traffic handled. But, despite the growth in air travel, it is still the old established inter-island shipping routes[8] which form the principal means of transportation for migrant Cebuanos. This statement applies with some force also to those still residing within the region, for the area's road systems, although above average by Philippine standards, usually offer speedy and efficient services only within the more heavily populated coastal areas.

The Western Visayas—the bitter sweet life

The Western Visayas produce two-thirds of all the sugar grown in the Philippines. The wealth thus generated, particularly from export earnings, makes for affluence without parallel in the Visayas. But, as is so often the case in the Philippines, areal and social maldistribution of wealth occurs on a large scale. Thus, for example, the life of the sugar barons of Bacolod in Negros Occidental Province is in vivid contrast to that of the salt pan workers of Roxas City in Capiz Province. These two elements of Western Visayan cultural geography epitomise the contrasts to be found in a region whose physical and cultural background exhibits a considerable degree of homogeneity. In 1970 an estimated 4·5 million persons inhabited the region's 21 600 square kilometres, totals which include the scattered islands of the Romblon group in the Sibuyan Sea (Fig. 32).

In both the island of Panay and the western portion of Negros Island, the land masses which account for the bulk of the region, the terrain consists of coastal lowlands extending well inland to the rugged hills and mountains which form the backdrop to almost every part. In Negros, the uplands are mainly volcanic in origin; in Panay, vulcanism has played a relatively small role in their evolution. The Panay highlands exceed 2000 metres in places and form a north-south spine whose eccentric positioning results in lowland areas being more extensive in the eastern provinces of Iloilo, Alkan and Capiz than in west facing Antique Province. In the province of Negros Occidental the lowlands are more extensive in its northern and central section. In the south the rolling Tablas Plateau is dominant although not in terms of elevation; the volcanic cordillera to the north towers to an elevation of 2438 metres at Mt. Canlaon. Throughout the region, the upland areas are clothed with near impenetrable forests. Dense stands of virgin rain forest are especially evident on the more inaccessible slopes of the volcanic cordillera in Negros; elsewhere, commercial logging and the *kainginero* have left their mark upon the landscape, giving rise to significant stretches of second growth forest and *cogon* grassland. Whereas in Negros lowland soils are derived from volcanic material, in Panay soils of alluvial origin predominate. Whatever their origin the region's soils are very fertile by Philippine standards. Drainage, however, can be a problem, particularly in Negros. Drainage problems become most acute when the dominance of the south-western air masses is most pronounced. In November-December through to May, in contrast, drought conditions are experienced in sheltered areas. However, areas exposed to the north and north-west experience heavy, year round, rains.

The peoples of the region share a common dialect—Hiligaynon—for during the Spanish era Negros Occidental was opened up by migrants from Panay, and common political and economic problems have further bound them together. Such problems generally have their roots in the product—sugar—around which life in the Western Visayas revolves. Commercial sugar cane production was introduced to the region in the nineteenth century; until then, the area was regarded by the Spanish as a rice surplus area supplying those less fortunate parts of the archipelago. Today, in contrast, the emphasis on commercial sugar production means that some places in the region, notably Negros, are food deficit areas; commercial sugar production has not only reduced the amount of land available for the production of foodstuffs but, by its labour demands, has also brought about spectacular increases in population.

One by-product of the population expansion resulting from sugar production has been a greater degree of urbanisation than in most parts of the Philippines. This is especially so in Negros where ease of communications and large scale sugar milling operations and associated manufacturing activity, such as the manufacture of sugar bags, encourage population concentrations (with a considerable share of the nation's factory workers, the island ranks second only to Luzon). Chief of these is Bacolod, the provincial capital of Negros Occidental; its 1970 urban population stood at an estimated 175 000 persons including most of the area's often extremely wealthy sugar cane planters who maintain palatial residences there and commute to their plantations when required. In addition to its normal functions as a provincial centre, Bacolod is noted also for the activities of farm implement and fertiliser dealers, shipping agencies and other middlemen all of whom are geared primarily to the sugar industry. The Planters Association and the Sugar Exchange, too, figure prominently in the life and livelihood of the city. Principal urban centres in Negros Occidental subsidiary to Bacolod include Silay (estimated 1970 population 42 500); Binalbagan, Victorias, La Carlota, Hinigaran and Talisay (all sugar milling centres with lighterage facilities and populations in excess of 10 000 persons in 1970); and Fabrica-Sagay which, in addition to sugar milling, also possesses one of the leading sawmills in the Philippines.

On the island of Panay urban dwellers do not comprise the massive 57% of the total population which makes Negros Occidental the second most urbanised province of the Philippines; nevertheless they form an above average portion of the total population. They are concentrated in Iloilo, capital of the province of the same name; in 1970 the city possessed an estimated 213 000 inhabitants and an influence which extends far beyond the provincial boundary, although, as an inter-regional trade centre, Iloilo's hinterland suffers from its location between those of Manila and Cebu. Although Bacolod today is perhaps the more dynamic centre, the traditional functions of Iloilo should persist for many years to come, thus ensuring that it remains the focus of the region's internal and external transport links with all the attendant implications in the realm of secondary and tertiary economic activity, and of culture and politics. In addition, furniture making and textile production have become established in and around the city; it supports several universities and colleges; while, to quote the Mayor, 'politics is a kind of industry hereabouts'. The capital of Capiz Province, Roxas City, with an estimated 1970 urban population of 27 500 persons, is the second largest town on Panay Island. Its port facilities, coupled with its position at the northern end of the trans-Panay rail route, has made it possible for the city to function as the service centre for a large portion of northern Panay. Other notable service centres on Panay Island include Estancia, New Washington, Kalibo and San Jose. Agricultural service centres located outside those areas where sugar is dominant handle a variety of products, the most important of which are rice (whose acreage throughout the region exceeds that of sugar), maize, *mungo* beans, root crops, tobacco and various fruits.

Non-agricultural pursuits rival farming in importance in parts of the region. Fishing is of special significance, the area being responsible for one quarter of the country's commercial catch. The population of the numerous villages which fringe the often low lying and swampy foreshores are partially or wholly dependent on the sea for their livelihood. Fishpond culture is also extensive, accounting for nearly one third of the nation's productive area; Panay Island, and the coast of Capiz Province in particular, are the locations where this activity is most widespread. Some mining takes place in the region with copper and gypsum the most significant. For both products mining is concentrated in the municipality of Sipalay on the edge of Tablas Plateau. Over 2000 persons are engaged in mining in the Western Visayas but even more are engaged in logging and sawmilling: activities are dominated by the complex at Fabrica but another score or so sawmills are operative also, mostly in Negros Occidental.

Life and livelihood in the Western Visayas is tolerable for many, agreeable for some and idyllic for a few. Except perhaps for certain marginal enterprises focussed upon Iloilo, economic prospects are bright. Furthermore, the region has strong political influence in Manila, and Bacolod is gaining a reputation as a miniature Manila in the field of entertainment. Yet in recent years the region has become an area of considerable population emigration. These migrants have not shared in the benefits accruing from the development of the region although they have made no small contribution to its progress. They are the labourers from the cane fields for whom life and livelihood have a permanently bitter taste. They are joined by their fellow countrymen from those other parts of the region where the cost of living may be low. In Capiz, for example, it is claimed by local officials that life is so cheap that fifty *centavos*—worth about five new pence—of fish could sustain a family of eight for a day, but the chances of making good are equally abysmal. For such people the sweet life of the Bacolod socialite is a symbol not of that which they might with ability and luck secure for themselves, but of the insuperable barriers created by man over the years as a result of maldistribution of natural resources and attendant socio-economic structures.

Eastern Mindanao—future heartland of the south

This region can be defined as encompassing the three provinces of Agusan, Surigao and Davao (Fig. 33). As such it occupies the eastern third (38 500 square kilometres) of the island of Mindanao. For a variety of reasons Mindanao differs substantially from other parts of the Philippines. Some of these differences are especially evident in Eastern Mindanao and none more so than the pioneer landscape created by migrants from other islands, particularly the Visayas, and their Japanese predecessors. In contrast to Central Mindanao, the pioneers in Eastern Mindanao had the advantage of a rather longer period of settlement, an injection of overseas (Japanese) capital and expertise, and an absence of ethnic and cultural conflict. The pioneer landscape therefore presents, in parts at least, a relatively mature appearance.

The dominant physical feature of Eastern Mindanao is the Davao-Agusan trough, a north-south downwarp or synclinal trough with a maximum width of forty-five kilometres. It is flanked on either side by structurally complex mountain ranges in the southern portions of which active vulcanism is an important element. The resultant instability of the physical features is accentuated by the region's susceptibility to earthquakes and, in the recent past, to uplift which caused the lowerlying portions of the Davao-Agusan trough to emerge from beneath the sea. The region's drainage is dominated by the north flowing Agusan River which on its route to the Visayan Sea incorporates a number of small lakes which, for the greater portion of the year, expand into extensive swampland areas. This phenomenon reaches its peak at the low sun period (November-May) when the bulk of the area's annual rainfall (in excess of 250 centimetres) is received. The more sheltered Davao Valley not only receives rather less rain (150-200 centimetres annually) but also possesses a more adequate natural drainage system. The drowned coastline formed where the region meets the Pacific contains the mouths of short and youthful streams which regularly carry away vast quantities of water, for annual rainfall here exceeds 500 centimetres in places. Significant stretches of the Davao-Agusan trough and the greater portion of the surrounding highlands are clothed with rich stands of tropical rain forest. Where clearance has taken place, soils with a relatively high organic content are revealed. To maintain their fertility, however, care needs to be taken to prevent leaching and erosion.

At the beginning of the twentieth century the population of Eastern Mindanao was concentrated in a handful of settlements. Their inhabitants had only limited success in the fields of agriculture and Christianisation of the indigenous population, the motives behind their establishment by the Spanish. It was left to the Japanese to call Eastern Mindanao to the attention of the new American Administration in Manila. The success of Japanese entrepreneurs was such that by the outbreak of the Second World War some 60 000 hectares in the southern half of the region had been opened up for the production of high quality abacá. In the absence of any marked interest in the region on the part of Filipinos the Japanese branched out into fishing, timber, and a variety of tertiary activities, notably retailing. This virtual outpost of the Japanese Empire was speedily dismantled at the end of the war and the void thus created was filled by the voluntary migration of Filipinos, particularly Cebuanos, whom the war had either uprooted from their previous abode or had given greater mobility and ambition. Unfortunately, this influx of people was not matched by an influx of expertise and capital. The resultant stagnation has not been entirely eradicated today but, increasingly over the years, an element of dynamism has been introduced. Building on the infrastructure provided by the Japanese, the initial

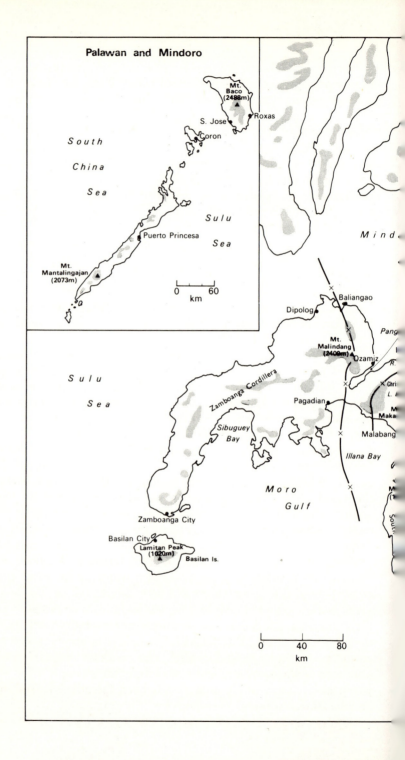

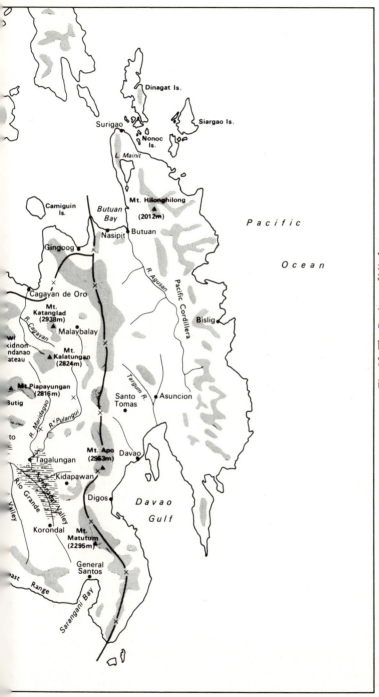

Fig. 33. The Regions of Mindanao.

haphazard development could hardly fail to progress, surrounded as it was by such a range of unexploited resources. In turn, development attracted more sophisticated entrepreneurs, both Chinese and Filipino, many of whom today control enterprises that bear comparison with any in the nation.

The base provided by the Japanese was confined to the southern portion of the region; elsewhere development had to start from scratch. It was spearheaded by forest industries which were supplemented later by agriculture, first orientated towards self-sufficiency but now increasingly operating on a commercial basis. At present, the latter is focussed upon coastal locations with their relative ease of access to markets in other parts of the Philippines, But, given the expected improvements in inland transportation following the completion of the Davao-Agusan highway, an upsurge in inland agricultural development can be expected.

Currently, maize production in the Davao hinterland provides the best example of how the agricultural potential of Eastern Mindanao has been realised. Here, some 250 000 hectares yield two and sometimes three crops yearly with yields about 50% above those experienced elsewhere in the Republic. High yields characteristic of virgin land are also being recorded by the abacá producers who have become established further to the north of Davao than previously in an effort to avoid infection from the abacá mosaic disease. Unfortunately, the development of maize and abacá production is not matched by other agricultural sectors in the region: Eastern Mindanao is a rice deficit area particularly in the south, where maize and abacá take precedence; production of coconuts and bananas on the other hand yields useful and growing surpluses but, as is the case with the region's other crops such as sweet potatoes, bananas and eggplants, quantities produced provide no true indication of the region's potential.

Timber extraction, an activity in which Chinese interests are to the forefront, has expanded rapidly in recent years to provide employment for half of the nation's forestry workers. In addition to satisfying a substantial local market, the products of the region's numerous and often impressive sawmills and wood processing plants supply many parts of the Philippines, and a thriving trade in logs with E.Asia and Europe exists also. Fishing makes a substantial contribution to the pattern of primary industry but resources are underdeveloped and generally operations are designed only to satisfy local demand. The region's mineral wealth is concentrated in its rugged eastern portion. Known resources include deposits of coal, iron ore, nickel, copper and gold, but only the latter has been mined in substantial quantities for any length of time. However, because of the lack of systematic exploration the possibilities for the discovery of further mineral deposits are high. Prospectors' optimism has been reinforced by the recently discovered laterite deposits in the Nonoc Islands. A minimum of 20 000 hectares are involved and feasibility studies by Canadian interests indicate that an annual nickel output of 20-25 million kilograms is possible. Furthermore, the iron content of the laterite (45·5%) has attracted Japanese interests who are considering the production of steel (eventually reaching 2 million tons per annum) from the laterite once it has passed through the nickel smelter.

Because of its pioneering nature, urbanisation has made little progress in Eastern Mindanao. Davao, with an estimated 1970 urban population of 150 000 persons, is by far and away the largest settlement. In addition to its inter-regional functions (especially those concerned with the export of abacá and the movement of regional surpluses of maize to deficit areas) it possesses, *inter alia*, veneer and plywood factories, a deepwater port for international shipping, a major hospital, and one of the country's leading hotels. Apart from Davao, settlements with a 1970 population in excess of 10 000 persons are few and far between. In the far north Surigao functions as a provincial capital and as a port with a hinterland extending southward well beyond the provincial boundary along the Pacific coast of the region. The capital of Agusan province, Butuan, is somewhat larger than Surigao, having an estimated 45 000 urban inhabitants in 1970. Its functions are similar to Surigao although both the area and potential of its hinterland are much greater. The other urban centres of the region function as agricultural service centres or mining and forestry company towns often with minor port facilities also. A few, such as Santo Tomas, have populations in the vicinity of 10 000 persons but numbers are more commonly in the vicinity of 5000. The greater number are located within the immediate hinterland of Davao.

Urbanisation is hampered by the imperfections of the region's transportation system. Physical barriers pose obvious problems on the margins of the Davao-Agusan trough but even here movement is possible through various breaks in the ranges. The recent completion of the Davao-Agusan highway marks a major step forward but much more needs to be done to provide the region's economic development with a sound infrastructure. Priorities in this field include the establishment of all-weather feeder roads and the hard surfacing of the intensively used Davao road network. Port facilities, too, are capable of substantial improvement. Major engineering work is also imperative in the Agusan Valley to combat seasonal flooding; not only would local transport be expedited but, more important, a substantial area of agricultural land could be opened up.

Between 1939 and 1960 about half a million persons are estimated to have migrated to Eastern Mindanao and its 1970 population stood at an estimated 2·1 million persons. For most of these people the region has provided employment opportunities unobtainable elsewhere and, more important, a promising future. It will be some time before these pioneers reap the full benefits of their efforts but indications are that the day is drawing steadily nearer when the region will cast off its pioneer image and emerge as the heartland of the southern Philippines.

Central Mindanao—the pioneer frontier

The near 50 000 sq. km of the provinces of Cotabato, Lanao del Sur and Bukidnon (Fig. 33) exhibit often markedly variable physical and cultural environments but they are united by the fact that, even more than in Eastern Mindanao, a pioneering atmosphere is apparent. In contrast however, the migrants to Central Mindanao found that they had been preceded in significant numbers by both pagan and Moslem (Moro) peoples. In Cotabato, the Moros pushed the pagans into the inaccessible Tiruray Highlands while they themselves occupied the extensive lowlands. Further north, the Moros penetrated the Lake Lanao basin but the rugged terrain deterred them from moving further east into Bukidnon. Moro settlement was essentially colonisation on the periphery of the Sri Vijaya Empire which lay to the south-west. It thus never really shed its pioneering nature and the region had to await the ending of the Second World War before determined attempts were made to exploit its potential. Substantial progress has been made in recent years but present day development in Central Mindanao falls short of that attained to the east as the Moros were unable to provide the impetus which the Japanese achieved in Eastern Mindanao and, more important, have created by their presence abrasive ethnic and cultural contrasts.

The dominant physiographic feature of the region is the Cotabato Valley. This intermontane lowland is half graben in structure and extends approximately eighty kilometres east and west and between eighty and 110 kilometres north and south. Recent uplift has impeded drainage with the result that the course of the Rio Grande and its major tributaries, the Maridagao and Pulungui Rivers, are frequently interrupted by extensive malarial swamps. This is especially the case at the Valley mouth where the Rio Grande forms two distributaries, the Cotabato and the Tamentaka Rivers, which in turn respectively traverse the Libungan Marsh and the Luguasan Swamp, the area's principal swamplands. The minimum area of the swampland totals 1150 square kilometres. The Valley is by no means a single physiographic unit as it consists of several synclinal basins one of which, the Korondal Valley, extends south-eastward to provide for the Valley a second exit to the sea at Sarangani Bay. Between Sarangani Bay and the mouth of the Rio Grande in Polloc Harbour, the coastline is backed by the horst block of the Tiruray Highlands and adjacent volcanic masses. These dissected uplands average fifty-five kilometres in width and achieve elevations approaching 1250 metres. To the north of the Cotabato Valley the terrain is plateau-like in nature. In its eastern portion, incision by the principal streams results in a mesa-type surface overlooked by numerous volcanic peaks, some exceeding 2750 metres in elevation. It was their eruption which covered the area with a thick layer of basalt whose upper surface lies between 600 and 900 metres. Volcanic action, too, created Lake Lanao (343 square kilometres), the focal point of the western portion of the plateau.

The climate of the region is characterised by constantly high temperatures and the absence of a dry season. However, the elevation of the northern portion of the region causes normal day temperatures only just to exceed 20°C. Rainfall in excess of 250 centimetres is experienced by most upland areas; normally, in the sheltered lowlands 125-200 centimetres are recorded. December-March is the period of least rainfall and in certain exceptionally sheltered areas, notably the head of Sarangani Bay, water shortages may result. Generally, however, the region's climatic conditions encourage the growth of dense tropical rain forest. Originally, the swamplands of the Cotabato Valley represented the only major non-forested area but, subsequently, *kaingin* agriculture, especially by the pagan Filipinos, resulted in the establishment of large areas of *cogon* grassland, notably in Bukidnon. With the advent of migrants from central and northern Philippines further areas have been felled for agriculture with the result that the forestlands now occupy less than half of the area of the region..

Detailed soil surveys have yet to be carried out but indications are that in terms of depth and chemical composition the region's soils compare very favourably with other areas of the Philippines. Where the parent rocks of the region are not volcanic in origin they are usually composed of limestone which also produces fertile soils. The main drawback would appear to be impeded drainage, a state of affairs which reaches sizeable proportions in the Cotabato Valley, the area upon which agricultural development is focused. Also, a marked fall in soil fertility results from the continuous cropping unless boosted by the regular application of fertiliser.

Hispanisation had little impact in Central Mindanao, the belligerent attitude of the Moros and the physiographic barriers to movement southward from the Visayas proving major inhibiting factors. The disappointing results of initial government sponsored settlement schemes beginning in 1913 further delayed the region's emergence as an integral part of the Philippines. Full scale migration did not get under way until after the Second World War but this marked the beginning of a flood which caused the region's population to soar from 0·7 million persons in 1948 to an estimated 2·3 million persons in 1970. This represented a growth rate of 229% which afforded striking comparison with the comparable national figure of just under 100%. The bulk of the migrants were attracted to the province of Cotabato which possessed an estimated 1970 population of 1·5 million persons (net migration to Cotabato Province in the period 1939-60 totalled 523 037 persons compared with only 231 445 persons for the rest of the region).

The influx of migrants who, with the exception of the relatively small numbers of Chinese, were predominantly Christians has created a situation fraught with cultural tension. The migrants not only brought with them a very different religious outlook from that of the indigenous peoples but also a markedly different concept of social and political behaviour. In view of the often radically opposed attitudes of the two sections of the population, a surprising amount of harmony resulted until 1970 when extremist groups resorted to violence in an effort to settle their differences. That major bloodshed was averted for so long was due in no small measure to the splitting in 1959 of the province of Lanao into a predominantly Christian Lanao del Norte and an overwhelmingly Moslem Lanao del Sur. This development avoided the creation of a political institution in which the opposing cultures would exist in uneasy balance. In Cotabato Province, the source of the extremist action, Moros and pagans together form a substantial proportion of the population. The latter, as in Bukidnon, form a largely passive element of the cultural environment but to some Moros, increasingly outnumbered by immigrants, violence appeared the only way to meet the threat to their traditional way of life, feelings doubtless exploited by local politicians and other vested interests, both Christian and Moslem alike.

Agriculture is the keynote of economic development in Central Mindanao. Mineral wealth is insignificant; inaccessibility at present precludes the development of the region's sizeable timber resources; and fishing, although possible along the coast, the rivers, the swamps, and Lake Lanao, remains essentially a sustenance operation despite significant potential for the establishment of fishponds. Maize and rice dominate farming in the region. Yields generally are not exceptionally high but output is well above local demand, so that considerable quantities can be shipped to food deficit areas, principally the Visayas. Root crops are grown on a large scale but these essentially pioneer crops are increasingly being supplemented by those with greater commercial potential such as abacá, pineapples, bananas, coffee, tobacco, jackfruit, mangoes, onions and Irish potatoes. The above mentioned vegetables are produced in the province of

Bukidnon where average elevations for agricultural land are at their highest. Here, too, the coconut is conspicuous by its absence. Pineapple production too makes a major contribution to the economy of this province for here, and in the Korondal Valley, leading American companies operate large lease-concessions, some exceeding 5000 hectares. This, and banana production, is currently experiencing major growth based on exports to Japan. Associated with pineapple production is the grazing of cattle based on the use of pineapple waste as supplementary feed. This is an activity which could well prove a significant pointer to future agricultural development especially for the *cogon* grasslands, which in Bukidnon in pre-war years supported a flourishing cattle raising industry.

With the region's economy orientated to agriculture at a pioneer stage of development, urban settlement in Central Mindanao provides the community with little more than its basic requirements. Cotabato and General Santos, each with a 1970 estimated urban population of 40 000 persons, are the only sizeable settlements. The former is the focal point of the region's transport network and, as a result, its service functions have a regional rather than a provincial flavour. However, its location on a low swamp-fringed interfluve between the two main distributaries of the Rio Grande would seem to act as a deterrent to the establishment of secondary industry and the more sophisticated tertiary activities necessary for it to emerge as a true regional centre. Settlements subsidiary to Cotabato all function primarily as agricultural service centres; the largest, with populations in the vicinity of 10 000 persons, include Kidapawan, Koronadal and Pagalungan. Even these are both primitive in services provided and distinctive in morphology: for example, settlement is not focused upon a Spanish church, as is typical in the Visayas and Luzon, but exhibits haphazard growth based on a single wide street straddling the highway and whose focal point is often the automotive service facilities.[9] Only General Santos can boast major port activities, for the region's coastline is mostly flanked by the South West Coast Range and thus lacks both good harbours and extensive hinterlands. The landlocked northern provinces look either to Cotabato or the settlements on Mindanao's northern coast for port facilities. In the northern provinces only their capitals—Marawi City (Lanao del Sur) with an estimated 30 000 urban inhabitants in 1970, and Malaybalay (Bukidnon) have populations in excess of 10 000 persons. The former could well emerge as a cultural centre of some significance for it contains the State University of Mindanao; it also has significant industrial potential because the nearby Agus River offers scope for the production of hydro-electricity.

Stimulus to urbanisation awaits improvements to the region's transportation network. Most roads are only dirt surfaced and many rivers must still be ferried. Inter-provincial road links are tenuous and swift passenger movement can only be achieved by regular air services which, however, are dependent upon inadequate airport facilities. A lack of feeder roads makes the local movement of both passengers and goods a laborious and hazardous business. The weaknesses of the road system are highlighted by the fact that riverine transport is extensively employed but, with streams usually only navigable by small craft, the movement of the principal items carried—bulk goods such as maize and rice—is far from efficient.

Today, the situation in Central Mindanao demonstrates that the brief hardships of the initial years of pioneering can be amply rewarded, especially as regards opportunity for individual landownership, and that the problems of malaria and coexistence with the Moros, although substantial, should not be insuperable. The ever increasing tide of migrants attracted to the region has reached such an extent that settlement subsidised by the National Land Authority now takes second place to that consisting of people who have migrated at their own expense. It will be some time before the pattern of life and livelihood is stabilised in Central Mindanao. This will have to await, *inter alia*, the establishment of an adequate transport system, agricultural practices geared to the local environment, the exploitation of timber and fishpond resources, and the provision of adequate urban facilities. The transformation of the region when this has been accomplished will not, however, lead to an environment akin to that of central and northern Philippines. By their act of migration, the settlers have in many ways put aside aspects of the traditional Filipino way of life; this, coupled with the unique features of the physical and indigenous cultural environment is creating in Central Mindanao a quite distinctive pattern of Filipino life and livelihood.

The Misamis Coastal Belt—an incipient industrial zone

The northern coast of Mindanao extending from Baliangao in the west to Gingoog Bay in the east (Fig. 33) represents both the area of densest settlement on the island (171 persons per square kilometre compared with sixty-eight persons for the island as a whole) and also that part most exposed to Hispanisation. To this has been added in the past decade the initiation of, by Filipino standards, substantial industrial activity. As a result, this comparatively small region (7000 square kilometres) has an estimated 1970 population of 1·2 million and exhibits a pattern of life and livelihood which sets it apart from its neighbours. Such is the smallness of the region that no province falls wholly within its boundaries but it does contain the bulk of the settled parts of Misamis Occidental, Lanao del Norte and Misamis Oriental.

The narrow strip of coastal lowlands and bordering band of foothills that make up the region seldom exceeds twenty-five kilometres in depth; indeed, the coastal plain itself averages less than five kilometres in width and is absent along many sections. The southern boundary of the region almost throughout consists of the escarpment of the Bukidnon-Lanao plateau. The foothills which flank the escarpment are composed mainly of Tertiary sediments; the lowlands have been built up from marine sedimentation, either recent deltaic deposition or older limestones, sandstones and the like. Only in the vicinity of the isthmus at the head of Panguil Bay do the lowlands attain substantial dimensions. The drawbacks posed by the lack of lowland areas and porosity of the limestone derived soils is accentuated by the region's climate. This is akin to that of the Central Visayas in that drought conditions prevail between February and April, and annual totals are comparatively low at 125-175 centimetres. Only on Camiguin Island and in areas lying adjacent to the plateau escarpment are good quality soils to be found, these having developed from the thorough decay of volcanic material. Unfortunately, such soils are usually associated with steep slopes and much of their value has been lost by soil erosion resulting from man's injudicious clearance of the vegetation. The dense tropical rain forest typical of the region remains untouched only in isolated parts. With agricultural land at a premium the amount of cleared land allowed to revert to second growth timber or *cogon* grassland is relatively small despite the steeply sloping nature of much of the land involved.

The accessibility of the region to the overpopulated lands of the Central Visayas and the similarity of the physical environment has attracted large scale migration over many decades. Today, people of Cebuano stock are in an overwhelming majority: even where they are the least dominant—in Lanao del Norte—they still make up four-fifths of the population. Throughout the region, the pioneering atmosphere of Central and Eastern Mindanao is generally absent: the evidence of Hispanisation, in terms of religion, town morphology, land tenure, etc., is as strong as in the regions to the north. The structure of the population provides further evidence of its maturity; an excessive youthfulness and predominance of males is absent from the Misamis Coastal Belt. Furthermore, with all available land occupied, population growth rates have largely stabilised.

Maize, rice and coconuts require the attention of the bulk of the rural population. The physical environment does not make for ease of cultivation and the region is fortunate in that the isthmus area, where the most abundant agricultural land is to be found, is exposed to air masses from the south-west which eliminates the drought season. This makes possible double cropping of rice and the creation of a maize surplus which is exported to the Visayas. Timber extraction makes a contribution to the rural scene but the bulk of such activity is concerned with logs secured from inland locations, beyond the region's boundaries, which are processed at the region's ports. The latter are also foci of fishing activity, but this is also a feature of the numerous villages which fringe the region's shores.

It is industry, however, to which increasing numbers of the population look for a living. The focal point is the city of Iligan which is strategically located to tap the hydroelectric potential of the nearby Agus River. It is estimated that the river is capable of producing three-quarters of a million kilowatts annually. Given that other less suitable sites can be developed along the courses of the other streams that plunge over the escarpment on their way to the Visayan Sea, the kilowattage potential of the region should exceed seven figures. At present, however, only 200 000 kW is developed from the Maria Christina Falls, but this is sufficient to have led to the establishment of an integrated steelworks employing 4000 persons with an annual capacity approaching 1 million tons (40% finished products, 20% each cold rolled sheets, hot rolled sheets and tinplate). In turn, the steelworks has encouraged the development of subsidiary construction and steel fabrication enterprises. The availability of hydroelectricity has also made possible the manufacture of cement, fertiliser, calcium carbide and ferroalloys. Unfortunately, the region is lacking in mineral wealth otherwise the pace of industrialisation would have been that much faster.

The industrial complex emerging in the vicinity of Iligan City caused its urban population to reach an estimated 45 000 persons in 1970. The largest settlement of the region, however, is still Cagayan de Oro. Cagayan itself had a 1970 population of about 55 000 persons and is surrounded by various satellite communities inhabited by a further 60 000 or so persons. In contrast to the 'boomtown' atmosphere of Iligan City, Cagayan de Oro presents a comparatively sophisticated appearance as befits a settlement which has emerged not only as the regional centre but also one with significant inter-regional functions. The latter revolve around the town's command of the major route link to the landlocked province of Bukidnon. Therefore, in addition to handling much of the commerce of Misamis Oriental (in which copra is paramount), Cagayan de Oro also acts as an entrepôt for large quantities of rice and maize. Fish products and manufactured goods comprise the bulk of its imports. A further manifestation of its links with the interior are the wharf and cannery facilities of the Bukidnon based Del Monte Pineapple Company. As a result of its commercial activities, Cagayan has emerged as the financial centre of the region and is also the focus of road, sea and air transport. Many other urban centres are to be found in the region, again reflecting a maturity of settlement exceptional for Mindanao. However, few possessed more than 10 000 persons in 1960 and most possessed less than 5000 persons. Most function as rural service centres and ports. Two settlements—Ozamis City (estimated 1970 urban population 25 000 persons) in the west and Gingoog (15 000 persons) in the east—function as commercial centres subsidiary to centrally located Cagayan de Oro. The importance of Gingoog is enhanced by its role as a logging and sawmilling centre.

The Misamis Coastal Belt stands apart from the rest of the island of Mindanao. Here, the potential of the island has reached an advanced stage of development to create a pattern of life and livelihood which has many counterparts elsewhere in the Republic. Industrialisation, however, is fast bidding to impose a unique character upon the region. The extent to which this affects the region as a whole will depend in part on the decentralisation of activity based on the hydroelectricity of the Agus River (a long term development as its potential is far from being fully tapped) but more upon the ability of Cagayan de Oro and other commercial centres to attract and encourage the establishment of less spectacular industrial enterprises. In attempting to do this they can point to the availability of labour in the region, a grid supply of electricity, the commercial and financial skills and contacts of local businessmen, and a potential market not just within the region itself, and the Iligan complex in particular, but encompassing Mindanao as a whole, and in particular the fast growing economy of the Lanao-Bukidnon plateau which lies within the hinterland of the region's ports. For many years it can be expected that such industrial development will be sporadic and mainly small in scale for at present in this sphere competition from Cebu is strong. But, ultimately, there would appear no reason why the Misamis Coastal Belt should not emerge as the industrial centre of the southern Philippines.

South-Western Mindanao—outpost of the Republic

The 250 kilometre long Zamboanga Peninsula disappears underwater at the Basilan Strait only to re-emerge as a 400 kilometre long chain of small islands linking Borneo to the Philippines (Fig. 33). The island chain, with the exception of Basilan Island, the largest of the group, falls within the province of Sulu. The remainder of the region is largely co-incidental with the two Zamboanga Provinces—del Norte and del Sur. In 1970 only an estimated 1·9 million persons inhabited the region's 18 685 square kilometres. In both physical and cultural terms, Sulu and the Zamboangas are dissimilar in many ways but they share a common marine orientation and feeling of remoteness from the mainstream of Filipino life and livelihood. The political storm which has arisen over the Philippines claim to Sabah has focused the nation's attention on the region as never before. In the long run, however, the region, paradoxically enough, could expect to receive the greatest attention from outside the Republic if its tourist potential is developed.

The south-western alignment of the region is a reflection of one of the main structural axes of the Philippines. The associated tectonic activity in folding the basement-complex rocks has encouraged volcanic activity but discouraged the establishment of low lying terrain. Vulcanism is especially evident in the far east, and on Basilan and Jolo Islands. In the first mentioned area, elevations in excess of 2250 metres are found although the divide of the Zamboanga Peninsula averages 600 metres. The islands vary in elevation from 1000 metres (Basilan Island) to islets a metre or two above sea level. Indeed, a characteristic feature of the island chain is its diversity of terrain resulting from wide variations in elevation, geology and shape.

The diversity of terrain is not matched by the climate which universally is hot and humid throughout the year. The well distributed rainfall varies from 125 centimetres in local rain shadow situations to over 250 centimetres in locations exposed to the south-west. In certain localities, however, water shortages may occur largely as a result of the porous nature of the soil. The soils involved are usually associated with coralline parent materials which have played an important part in the evolution of the Sulu Islands, and the tip of the Zamboanga Peninsula. Other soils found in the region generally are developed from volcanic materials, in which case their fertility is above average, or else from the basement-complex rocks, in which case the opposite applies. The vegetation pattern is responsive in detail to soil conditions but the overall presence of tropical rain forest reflects the dominating influence of climate. Deforestation is widespread, especially on the islands and more particularly where soils of a volcanic origin are present. Substantial areas of virgin forest remain, however, principally on the Zamboanga Peninsula but also on certain of the islands, notably Tawi Tawi. They are characterised by an exceptionally high volume of merchantable timber per acre.

The population of the region falls into two distinct categories—the Christianised Filipinos of the mainland who have migrated there during and since the Spanish era and whose homeland generally was the Central Visayas, and the Moslem seafaring peoples of Basilan Island and Sulu who moved into the region from the south-west during and before the Spanish era. The latter are to be found only in this part of the Philippines; the Moros of mainland Mindanao have close affinities with them but they have largely forsaken the sea as the focus of their life and livelihood. Many individual ethnic groups can be distinguished amongst the seafaring peoples but it is sufficient to single out the Tausog and the Samal Laut. The former prefer a littoral environment, living in pile built dwellings and engaging in both agriculture and fishing (not to mention legal and illegal trading, and even piracy). Although somewhat diluted by intermarriage with other ethnic groups, both seafaring and non seafaring, they are still noted for their belligerent attitude, religious fanaticism and ethnic pride. Jolo and Basilan Islands are the principal areas in which they have settled. The Samal Laut, on the other hand, are primarily or exclusively dependent upon the sea and are to be found throughout the islands. Although the degree of mobility varies, the term 'sea gypsies' is not inappropriate, for such people are characterised by a preference for a nomadic life spent at sea in groups consisting of related or friendly families. In the past, piracy and trading (that with Sabah being of special significance) were the keynote of life and livelihood but with improved law enforcement greater attention has been paid to fishing and home bases have been established. Nevertheless, there remains an ingrained opposition to any attempt to impose restrictive elements of modern civilisation such as custom controls and allegiance to political parties.

The marine orientation of the population is exemplified by conditions in the province of Sulu. Here, agricultural land use (devoted primarily to coconuts, rice, abacá and root crops, especially cassava) takes second place to fishing and associated maritime enterprises. Among the latter, the smuggling of cigarettes and other contraband from Sabah certainly receives the most publicity although the wealth generated probably is by no means equally distributed. Less spectacular but more fundamental is sponge gathering for the subsequent processing into bêche-de-mer; shells (for buttons and mother of pearl), pearls, black coral, clams (for dried meat) and turtles (for meat and eggs) are equally assiduously sought after. Fishing *per se* is focussed upon sardines, anchovies, mackerel and tuna. Unlike the more exotic products described above, the market served is primarily local in nature. Commercial fishing is better developed in the Zamboangas which account for 2% of the nation's commercial catch. Agriculture, both on a commercial and subsistence basis, is also better developed on the mainland and on Basilan Island. Coconut cultivation is a major occupation and output accounts for 8% of the nation's copra exports. Abacá and rubber also make a contribution to the agricultural land use pattern, despite a lack of large scale capital investment. Food crops produced include rice, maize and camotes. Maize is produced at a level of output in excess of local needs but the region as a whole, and Zamboanga del Norte in particular, is a rice deficit area.

Both logging and minerals contribute to the regional economy. The former is focussed on Basilan Island from which some 8% of the Philippine timber cut originates. Mining is confined to the mainland where the principal attraction is coal and iron ore deposits located in the centre of Zamboanga del Sur. The coal mined is moved to the Iligan steelworks but of more importance are the open pit iron ore mines, the output of which is exported to Japan.

Transport throughout the region is orientated to the sea. Roads are of negligible importance on the islands and those on the mainland are usually inadequate in construction. Every settlement of note is therefore a port. Zamboanga City is the focus of inter-island movement and is also the principal port of call for inter-regional shipping. No industrial activity of substance has been attracted to the port, so that without an administrative role (Dipolog for del Norte and Pagadian for del Sur being the provincial capitals) and with an immediate hinterland which is agriculturally unproductive, the port is devoted almost exclusively to commerce (it is the fifth ranking inter-regional trade centre, being particularly concerned with inbound rice shipments and the collection and forwarding of local copra and fish). The urban population of Zamboanga City in 1970 was an estimated 40 000 persons; the second largest settlement on the mainland—Pagadian—is little more than one half as large. The latter commands the largest area of agricultural land on the mainland and acts as an administrative centre, agricultural service centre and port. Dipolog (20 000 persons in 1970) functions in a similar manner, eighty-five kilometres due north of Pagadian. Dipolog, however, has no pioneering atmosphere to contend with (the Pagadian hinterland has been opened up only in the past decade). Other settlements with populations in excess of 10 000 persons in 1970 consisted of Basilan City (14 000 persons) and the only urban settlement of any size in Sulu Province—Jolo (40 000 persons).

South-Western Mindanao is not just an ethnic and political outpost. Living standards compare unfavourably with other parts of the Republic. Thus, for example, the frequency of home radios, piped water, electric light and other domestic facilities is amongst the lowest recorded in the Philippines. The differing way of life in the islands is the prime cause of this state of affairs but it should also be noted that the two Zamboanga provinces are amongst the least wealthy in the land. For the mainland, hopes for improvement must be focussed on its timber and mineral wealth; for the islands, tourism could provide even greater rewards. The latter development would be all the more welcome if anti-smuggling measures eventually prove successful, for the other mainstay of the island economy (marine products) has suffered a decline in recent years as a result of the cutting off of one of its main markets—Communist China—plus growing world competition and the increasing use of plastic substitutes. Success in promoting tourism is by no means certain. The real or imagined fear of the Moro on the part of the average Filipino will only be overcome with patience, and large scale capital investment will be essential to divert the international tourist en route to Hong Kong, Bangkok and other established tourist centres of South-East Asia. But the main obstacle to the development of tourism lies in the people themselves—they have created a unique setting but individual temperaments will have to be held severely in check if the region is to reap any material benefit. For people like the Samal Laut material gains may be offset by non-material losses, but, regrettable as this may be, accelerated economic development appears highly desirable for this remote part of the Philippines.

Palawan and Mindoro—the empty islands

The islands of Palawan and Mindoro (see Fig. 33) make a substantial contribution to the total land surface of the archipelago (25 140 square kilometres or 8·4%). Despite this they are inhabited by only an estimated two-thirds of a million or about 4% of the national 1970 population. In the case of Palawan, isolation is an obvious causal factor and, on closer examination, the nature of the physical environment is seen to act as a substantial deterrent also. The reason in the case of Mindoro is less obvious but it would appear that its malarial and densely forested lowlands and inaccessible mountainous interior led to its being bypassed despite its location athwart the main sea routes between Luzon and the Visayas. To would-be migrants, Mindanao not Mindoro was the lure but recently, with Mindanao approaching saturation point, settlement on Mindoro has accelerated in a modest fashion. Mindoro could well shrug off its past neglect in future decades but for the present it and Palawan together consitute a physical, economic and political backwater.

As regards physical conditions, the islands have strong associations with Borneo. They represent exposed outliers of the Sunda Shelf and consist of an ancient land surface in which vulcanism is conspicuous by its absence. Palawan is noted for its rugged terrain which is associated with a mature karst topography which has exposed the underlying basement-complex rocks. The highest elevation exceeds 2000 metres. The basic ingredients of the Mindoro terrain have been mentioned above. Interior elevations can exceed 2500 metres and, as in Palawan, basement-complex rocks (granites, gabbros, schists, etc.) predominate, being exposed by erosion of the overlying thick limestone deposits. Unlike Palawan, erosion has given rise to extensive lowland areas, particularly in the island's eastern half, which contrasts sharply with the rugged nature of the interior.

The differing alignment of the two islands causes them to experience distinctive climatic conditions. In Palawan, its southwest to north-east orientation means that the island lies parallel to the movement of the dominant air masses. Drought periods are thus heightened and although rainfall totals generally exceed 175 centimetres this often is inadequate in view of its marked seasonal distribution and the nature of the soil. In Mindoro soil conditions are basically similar but, as the island's alignment is at right angles to that of Palawan, two contrasting rainfall regimes are apparent. For Western Mindoro amounts in excess of 250 centimetres are received mostly between June and September whereas in the eastern half of the island the north-east monsoon brings the bulk of the annual rainfall of 200 to 250 centimetres during the 'autumn' and 'winter' months. The general absence of any dry season in the east is offset to some extent by the fact that it is rather more prone to typhoons.

Poor quality soils are characteristic of the basement-complex rocks while those which have limestone as a parent material, although rich in calcium and phosphorus, suffer from excessive porosity. Nonetheless, climatic conditions are such as to ensure a dense forest cover which has close associations with that of Borneo. Typical species are dominated by extreme hardwood species and include *apitang, molave, ipil* and *narra*. In Palawan some 70% of the forest cover is untouched; in Mindoro, *cogon* grassland and secondary forest are more extensive while mangrove forest is to be found along parts of the coast.

Population densities on the two islands, at fifteen persons per square kilometre (Palawan) and forty persons (Mindoro), are amongst the lowest in the Philippines. The Spanish made determined efforts to colonise Palawan as a base for anti-Moro activities but found physical conditions too great an obstacle. Settlement in Mindoro is of more recent origin and at present is focussed upon small agricultural villages scattered along the coast. The largest of these settlements rarely contain more than a few thousand persons. On Palawan, easily the largest settlement is the capital, Puerto Princesa, with an estimated 7500 persons in 1970.

Primary industry is dominant throughout the region, the only exceptions of significance being the commercial activity in the larger settlements, a sugar mill in Mindoro, and several sawmills, the most important of which are to be found in Mindoro. The greater proportion of the labour force is engaged in agricultural pursuits principally involving coconuts and rice, and also sugar cane in Mindoro. On Palawan the lack of suitable alluvial soil results in an above average emphasis on upland rice. With both the Moslem inhabitants of Southern Palawan and the pagans of its interior engaged in *kaingin* agriculture, this results in the island's population placing an exceptional reliance on sloping land for their food supply.

Fishing is a close second to agriculture in certain parts of the region. Subsistence fishing is universal, for most settlements have a coastal location. Commercial fishing also plays an important role, especially in the waters surrounding Palawan which supply two-thirds of Manila's fish supply and one sixth of the nation's commercial catch. The nation also looks to the region for much of its salt, there being numerous solar evaporation pans in Western Mindoro which produce 30% of the country's salt. Other mineral wealth worked mainly comprises deposits of manganese, mercury, chromite and silica sand on Palawan Island.

The mineral wealth of the region, and also its marine and timber resources, have yet to be exploited to the full. Palawan in particular will have to rely heavily on non-agricultural resources in the future; in contrast, the opening up of the Mindoro lowlands by immigrants should result in that island possessing a modest but flourishing agricultural economy in the years to come. Modest progress probably will be the keynote of the region's development. With the exception of commercial fishing, which in many instances is Manila based, available resources are not large compared with those to be found elsewhere in the Philippines. Migration should bring Mindoro more into the mainstream of Filipino life but both it and Palawan will remain, if not neglected, at best secondary components of the geography of the Philippines.

References

1. J.E. Spencer, *op. cit.,* pp. 35-6
2. Individual geographical studies of specific regions of the Philippines have been few in number. They include the following:
   T.W. Luna, Jnr., 'Some aspects of Agricultural Development and Settlement in Basilan Island, Southern Philippines', *Pacific Viewpoint*, Vol. 4, 1963, pp. 17-24.
   M. McIntyre, 'Geographical Regionalism in Leyte', *Philippine Geographical Journal*, Vol. 3 No. 1, January-March 1955, pp. 31-9.
   M. McIntyre, 'Geographical Regionalism in Samar', *Philippine Geographical Journal* Vol. 4 No. 1, January-March 1956, pp. 7-13.
   F.L. Wernstedt and J.E. Spencer, *The Philippine Island World: A Physical, Cultural and Regional Geography,* University of California Press, Berkely and Los Angeles, 1967. It is upon this work that much of this chapter inevitably is based.
3. It should be noted that normally the designated boundaries of Philippine cities lie far beyond the limits of urban development. Often considerable numbers of rural inhabitants are enumerated with urban dwellers in arriving at population totals. In this chapter every effort has been made to eliminate non-urban inhabitants, particularly where the larger settlements are concerned.
4. B.L. Starner, 'Report from Arayat', *Far Eastern Economic Review*, Oct. 15th 1966, p. 147.
5. F.L. Wernstedt and J.E. Spencer, *op. cit.,* pp. 277-8.

6. T.W. Luna, Jnr., 'Manufacturing in Greater Manila', *Philippine Geographical Journal* Vol. 5 Nos. 3-4, July-December 1964, p. 81.
7. F.L. Wernstedt and J.E. Spencer, *op.cit.*, pp. 400-1.
8. F.L. Wernstedt, 'Cebu, Focus of Philippine Inter-island Trade', *Economic Geography*, Vol. 32, 1956, pp. 336-46.
9. F.L. Wernstedt and J.E. Spencer, *op.cit.*, pp. 555.

7
The Past

The advent of the Spanish in the sixteenth century A.D., and the almost coincident arrival of Islam in the southern Philippines, radically modified the details of earlier patterns of Filipino life and livelihood. However, in no small manner the basis of this pattern survived to make its contribution to the present day situation. Therefore, it is necessary to consider the state of affairs which existed in the Philippines prior to the arrival of Magellan.

The Pre-Spanish Era

The first immigrants to the archipelago probably arrived in pre Neolithic times prior to its severance from the mainland of South-East Asia, and later, from the neighbouring island of Borneo. Despite the resultant ease of penetration, the Negrito immigrants and their proto-Malay successors were small in numbers; this fact, coupled with an economy little advanced from the hunting and gathering stage, meant that their modification of the physical environment was of negligible proportions. More probably, it was volcanic activity and associated tectonic movements which contributed most to the evolution of the present day geography of the Philippines in this period. However, starting several thousand years before the birth of Christ, the initial inhabitants of the Philippines were joined by Neolithic peoples who exhibited basic mongoloid racial features and, by the birth of Christ, yet another racial group, the Malay Mongoloids (genetically very similar to the proto-Malays), were making their presence felt. Their arrival heralded an era in which modifications of the environment by nature, although often violent, were to become of less importance as man, through a combination of population pressure and improved techniques, increasingly sought to impose his will upon the archipelago.

By now, the Negritos and proto-Malays were experiencing the competition for land which increasingly was to delimit their sphere of influence. Over the years, intermixing has resulted in the assimilation of the more primitive peoples but for many a retreat to ever more inaccessible areas provided the only means by which they could pursue the way of life and livelihood to which they were accustomed. Nowhere in the Philippines are Negritos and proto-Malays sufficiently important in terms of population or area occupied to play a dominant role in its regional geography as defined by this study. Yet, they are present on all the major islands, notably Luzon (particularly Mountain Province), Mindanao (Bukidnon-Lanao plateau), Palawan, Mindoro, Panay and Negros. More significant, these peoples' activities possess features which are often the only current expression, albeit distorted, of past behaviour upon which the present day life and livelihood of many Filipinos is founded.

The practice of *kaingin* agriculture is an outstanding example. Most Negritos and proto-Malays are *kaingineros*; pure hunters and gatherers are rarely found. Typical of the former are the pagan group who eke a living from the rugged forested terrain of the interior of Southern Negros.[1] Popularly known as Bukidnons, their racial origin is probably proto-Malay. They produce upland rice, maize and root crops by means of *kaingin* agriculture and supplement these crops by naturally occurring edible roots, fruits, and vegetables. Wild life provides the only indigenous protein source; this may be supplemented by fish obtained through intermittent barter trade with the inhabitants of the lowlands. Isolation, and resultant ignorance, coupled with restriction of movement due to logging activities, the debilitating effect of the rigorous physical conditions, and inadequacies of diet make the Bukidnon incapable of employing *kaingin* agriculture to the maximum advantage, as a long range system of land rotation. Instead, too drastic clearance and too frequent return to previous sites lowers productivity and permanently modifies the forest ecology. In Southern Negros the population-land ratio as yet is not critical but in certain parts of the Philippines, notably the Bukidnon-Lanao plateau and

Mountain Province, population pressure and/or a commercial incentive resulting from either the establishment of regular barter trading or the practice of *kaingin* agriculture as a supplementary source of income by more sophisticated Filipinos has resulted in the permanent establishment of large areas of second growth forest or *cogon* grassland.

The way of life of the Negrito and proto-Malay is unique. It is of particular interest as certain features provide the only evidence extant of the cultural pattern which existed at the arrival of the Spanish. The Isneg, one of the various groups of aborigines whose home is the mountains of Northern Luzon, can serve as an illustration.[2] They are best known for their headhunting activities but of greater relevance to this study are features such as their language, religious beliefs, and kinship ties. Nevertheless, it should not be forgotten that the persistence of headhunting amongst the Isneg is another relic, in an extreme form, of the traditional penchant of the Filipino for violence. Headhunting, too, has a ritualistic aspect and the Spanish were able to substitute the ceremonial and pageantry of Roman Catholicism for the varied and often bloodthirsty rituals practised by the Filipinos in the veneration of *anitos*, or spirits, The importance attached to such veneration is in large measure a consequence of the focusing of hopes and anxieties upon the annual crop cycle and related phenomena, as well as aspects of sickness and health. The numerous manifestations of these basic determinants of primitive society led to the creation of a profusion of malign and benificent spirits (which in the case of the Isneg run into several hundreds) for whom the Christian saints proved to be an acceptable substitute.

Perhaps the most potent features of Isneg culture are family and kinship for these closely parallel those currently exhibited by the average Filipino. Multiple family households are normal, and associated kinships mean that the average Isneg might have about one hundred relatives close enough to be taken actively into account in appropriate social situations. In the case of the Isneg, these interwoven relationships are heightened by intermarriage, the physical difficulty of contacts with neighbours and the barrier formed by the distinctiveness of their language, itself a product of isolation. Only in recent years has such parochialism been broken down in any significant degree as a consequence of the impact of a more mobile, modern civilisation. The Isneg, however, have been successful in retaining distinctive styles of clothing and implements, material features of the more accessible rural areas which were rapidly to undergo considerable standardisation as a consequence of Spanish colonisation. Finally, it should be noted that the precariousness of male survival among the Isneg has led to the female playing an important role in economic and cultural affairs. This is a phenomenon characteristic of much of Filipino society: the Spanish were unable to eradicate it even though it clashed with their own cultural heritage, and subsequently the Americans, to whom it struck a more sympathetic note, encouraged it.

The Neolithic and post-Neolithic migrants to the Philippines represent the most important element of the present day racial mix, accounting for a 40% share. By the arrival of the Spanish they had spread throughout the islands and occupied most of the more favourable terrain. These people brought to the Philippines the basis of current boat design; wooden and bamboo house styles; fishing, pottery, mat and basketwork techniques; clothing styles; and a great many domesticated plants and animals, and farming customs. The more sophisticated of these innovations were Indian in origin and in the pre-Spanish period cultural penetration of this nature left a permanent mark in such diverse fields as language, art motifs, agriculture, architecture, and political organisation. Such influences spread into the Philippines from the south-west, the most recent being brought by the traders, missionaries, and political organisers who had succeeded the migrants and who were based on the Sri Vijaya and associated Empires of the Indonesian archipelago.

In the years immediately prior to the arrival of the Spaniards, however, a countervailing influence was waxing strong. This was Chinese in origin and was felt most markedly in Luzon. In this case innovations were more commercial than cultural and involved greater direct contact. The latter resulted in an influx of people, mainly from Southern China who, by assimilation over the years, have contributed a 10% share to the present mix.

By the close of the fifteenth century A.D., therefore, the pattern, if not the density of the population, bore marked resemblances to that of the present day. Similarly, many of the present day crops were in production, cultivated by people whose racial features, life and livelihood in type, although not in manner, exhibited many characteristics recognisable today. The basic framework of the country's present day geography was thus established. The changes which were to take place in the ensuing centuries, spearheaded by the introduction of Christianity and Islam, were to have an increasingly profound effect. However, except in limited circumstances, they have not destroyed beyond recognition the traditional way of life and livelihood of the Filipino.

The Spanish Era

The religious influences which reached the Philippines at this time had wide social implications. Existing amorphous cultural patterns became institutionalised and many of the present day regional and national patterns became firmly established. Receptiveness was the keynote of Filipino culture at this time: Spanish institutions in particular were successfully adjusted to Filipino values while at the same time large sectors of basic Filipino culture were retained. The result is a unique combination of Oriental and Occidental cultures.

It was not until 1565, when Legaspi established his headquarters on Cebu, that the spread of Islam was successfully challenged. By this time, it had been present in the Philippines for some eighty years; yet, despite a late start it was Roman Catholicism that, in a few decades, had established itself in the eyes of the greater portion of the population as the 'true' faith. In no small measure the steady progress towards conversion to Roman Catholicism owed its success to the way in which the friars exploited the Church's ability to serve as an acceptable substitute for the pagan rituals then in vogue. A major contributory factor was the lack of success experienced by the Spanish in their economic objectives, which normally took precedence over religious matters. Neither precious metals nor spices were available in any quantity so conversion of the heathens provided some compensation to Mexico for the heavy subsidies it continually supplied to prop up the Filipino economy.

Islam, therefore, had to be content with a sphere of influence which only encompassed the extreme south and south-west.

Roman Catholicism was unsuccessful in penetrating this section of the Philippines in part because of the physical barrier posed by the mountains that pressed close upon the northern shores of Mindanao; in part by Moro (Moslem) control of the Sulu Sea and adjacent waters; and in part because of the time factor which had allowed the build up of a Moslem tradition amongst people in whose more aggressive outlook the religion found greater sympathy. It must not be forgotten that, by the arrival of Legaspi, Islam was practised as far north as Manila. However, the nature of its spread proved its downfall. North of Mindanao, adherents to Islam were predominantly *barangay* leaders originating from the Moslem core area, who often had usurped this role by force as a consequence of piratical incursions, and who consequently had a weak internal power base and few allies with which to resist Spanish penetration.

But although Islam was but a passing phenomenon, the political ideas that accompanied it did not wither away. Moslem leadership effectively formalised the political structure of the *barangay*, notably by defining the role of the headman or *datu*. This evolution of the *barangay* from a kinship unit (usually comprising thirty to one hundred families) to an economic and political unit (the *barrio*) was speedily completed by the Spanish. The *barangay* had its territorial limits defined and the *datu* ceased to be an hereditary appointment. The *teniente* was the new community leader who, although retaining some of the *datu*'s powers, became the instrument through which the central administration influenced local affairs and, more important, collected the taxes. Today, the *teniente* and his immediate superior (the *poblacion* mayor) perform much the same function (except insofar as taxes are concerned) while the local priest also remains as 'the power behind the throne'.

The *barrio* and its smaller version, the *sitio*, formed the lowest rung of the administrative (and religious) ladder[3] set up by the Spanish which remains operative today.[4] More important *barangays* evolved into *poblaciones* which in turn served as administrative centres for the *municipios* (known today as municipalities) as broader groupings of *barangays* came to be called. On a yet broader base, which in the Visayas might comprise a whole island, provinces were established but these, too, were subservient to the central government. Officials at all levels were not elected by the democratic process as is the case today; instead, political and social partronage were all-important, taking precedence over merit or popularity with the masses.

The appointive power of the Church, however, is now considerably muted for the formal link between it and the state which characterised the Spanish era was speedily broken at the end of the colonial period. Nonetheless, involvement of the Catholic hierarchy in politics and government is still an accepted fact. In Moslem dominated areas, and in remote areas in particular, this state of affairs is even more prevalent for here political decisions have little hope of succeeding without prior approval by the religious leaders. Their power of arbitration over community matters is as wide as that experienced by the parish priest in his heyday.

The political hierarchy established by the Spanish also set the pattern for the structure of society. Power rested with the administration, the Church and the military, and as a result their leaders became the leaders of society. By definition also this upper strata, particularly above the *municipio* level, was almost exclusively Spanish. Gradually, Filipinos gained entry, chiefly as a consequence of status acquired by land ownership. Today, land ownership is but one of many status symbols but in large measure the upper echelons of society still can be most easily penetrated by those persons who were successful in the past— the landowners, the politicians, the military and the priests.

The social status attached to land ownership in the Philippines has had many undesirable consequences. Its implications for agrarian reform are discussed below but at this point it is worth examining its effect on economic growth in general. Land ownership was coveted for economic as well as social reasons and, of the former, none was more compelling than the fact that in an unstable country such as the Philippines land was the most secure of assets. The wealth derived from landownership in the Spanish era was thus in many instances used to acquire more land to the subsequent detriment of other sectors of the economy. More important, despite increased stability in the post-Spanish period, traditional attitudes have been slow to change. Thus, not only is agricultural reform thwarted but a low level of investment in industry and commerce has led to a restriction of growth in these sectors of the economy and their falling into non-Filipino hands.

Another detrimental tradition established in the Spanish era, and linked to the obsession with landownership, has been the stifling of initiative on the part of the population at large. This state of affairs is also a reflection, *inter alia*, of a feeling of inferiority created by the American presence and the success of the Chinese in penetrating most decision making areas of industry and commerce. But fundamental to the Filipino attitude of letting someone else think for him when major decisions have to be made is the peasant mentality, ingrained as a result of pre-Spanish social organisation and formalised and accentuated by Spanish landholding practices.

Spanish practice from the first was to recognise native rights to such lands as were actually being occupied and cropped (the *datu* and his nobles who possessed these rights thus obtained formal, though not legal, recognition of their status and were henceforth known as *caciques*). All other land, however, became Crown property. With the steady growth in population in the Spanish era (from approximately half a million to some 7 million) large areas of such Crown land were subsequently occupied. Crown lands thus affected were disbursed by the government to persons of influence, usually tribal leaders and upper class families in the case of Filipinos, or those to whom it was indebted, notably pensioned Spanish soldiery. Other land was held by the Spaniards under the *encomienda* system. Such land in theory, though not always in practice, reverted to the Crown but what did persist was the practice associated with the *encomienda* system. Thus was perpetuated the subservient position of the bulk of the population and the hereditary preeminence enjoyed by the few, for the *encomienda* system paralleled semi-feudal practices of the *barangay* in that land grantees (*encomienderos*, later known as *hacienderos*) had the right to collect a tribute from the produce of the land and to exact labour service from its inhabitants. Thus, the *datu's* serfs or slaves of the pre-Spanish era became the share tenants (*kasama*) of the *cacique*.

The *encomienda* system differed from pre-Spanish land occupation in that usually much larger holdings were involved. Furthermore, in the latter part of the Spanish era *hacienderos*, by virtue of their better education, were able to expand the size and make permanent the tenure, of their holdings through exploiting the Royal decrees concerning landownership registration. In this way, too, adjacent *caciques* were dispossessed by force or by trickery of their rightful heritage. Another abuse of the system involved exploitation of the peasantry. In this respect the so called 'friar lands' came to exemplify the situation. The Church, in the form of the religious orders of the Dominicans, Agustinians and Recoletos, was able to acquire, through its intimate connection with the government, some of the best agricultural land in Luzon. Despite relatively high productivity, therefore, the *kasama* saw the cream of his output appropriated by owners whose interest in agriculture was minimal and whose demands for labour services were directed towards the creation of works designed for the glorification of God rather than the material benefit of the *kasama* and his family.

Other agricultural developments in the Spanish era generally did little to ameliorate the conditions under which the peasantry earned a living. Few improvements in technique were effective for not only were the Spanish by nature disinterested in agriculture but the tropical conditions under which it was pursued were foreign to them.

Changes which did occur in agriculture and which are of relevance today focused upon the introduction of new crops and animals, chiefly American in origin. Of these, maize, tobacco, pineapples and horses in time were to play an important role. Tobacco was of special significance. Until the second half of the eighteenth century agriculture in the Philippines was almost entirely one of subsistence. Each region—one might almost say each settlement—produced what it needed for its own consumption. What trade there was consisted chiefly in supplying farm products to the non-primary producers of Manila and the larger towns in exchange for imported manufactures. The tobacco monopoly, however, by introducing agricultural specialisation on a significant scale, created a demand for staples, such as rice, and hence stimulated their production for the market.

Maize, like tobacco, was planted soon after the arrival of the Spanish but large scale production did not take place for some two hundred years. As Vandermeer[5] points out, conservative attitudes (local farmers traditionally grew native millet and, in addition, disliked the taste of maize) coupled with repeated confiscation of crops by government officials delayed its introduction on a large scale until population pressure left no choice but to switch to the higher yielding grain.

Notable by its absence was systematic irrigation. The pre-Spanish inhabitants initially avoided water deficit areas. When forced by population pressure to occupy such areas they lacked

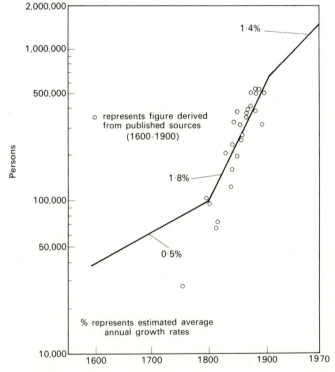

Fig. 34. Population Growth in Cebu Province 1600–1970. In the twentieth century the growth rate has exhibited a slowing down due primarily to outmigration

the skill to establish anything more than haphazard schemes. The arrival of the Spanish did not alter the picture for they lacked practical experience, and, equally important, funds with which to construct and maintain appropriate irrigation systems on a large scale.

It was not until the nineteenth century that the rural scene fully reflected the innovations of the Spanish (and the Chinese). More stable political conditions in particular stimulated commercial agriculture. In those areas most intimately concerned the abandonment of near subsistence agriculture had a dramatic effect on population growth (Fig. 34). Drought, typhoon, disease, etc., remained, but the diminution of Moro depredations, the end of the Hispano-Dutch war, and the opening of Manila to world commerce in 1834 brought a modicum of prosperity, which for the common *tao* primarily meant a longer life span in which to raise more children.

It was in the urban areas that material changes resulting from Spanish rule were most evident. Perhaps most spectacular were the fortifications, remains of which form important tourist attractions today and demonstrate the importance attached in the past to protection against other colonial powers, internal rebellion and raids by Moro pirates. But, from the point of view of Philippines urban geography, the Spanish imprint on the settlement pattern is of more consequence. This was highlighted by a grid line arrangement of streets oriented to the *plaza* around which were grouped the church, the treasury, the headquarters of the bureaucracy, and the garrison house. To these manifestations of Spanish domination were added the homes of the leading local dignitaries. This rigidity of the Spanish settlement pattern exemplifies the institutionalisation of society that took place at this time.

Assimilation of Spanish culture in terms of language, clothing and housing styles, education, etc., was far from universal and rarely of substance. Thus, the most typical effect would be modifications to food habits or the incorporation of Spanish words and phrases into the local dialect. The permeation of these aspects of Spanish culture was most effective where intermarriage took place, a not infrequent occurrence, as few Spanish women went out to the Philippines. However, compared with unions entered into by the Chinese, the overall effect was limited for at any one time there were usually no more than some 2000 Spanish officials present in the country. Individual exposure to Spanish culture being thus limited, current attitudes cause it to be regarded as something which, while sophisticated and deserving of respect, is irrelevant to the present situation.

Vividly contrasting with the indifferent attitudes to Spanish cultural patterns has been the response to the civil and military bureaucracy set up by the colonial government. Resentment towards its inefficiency proved to be a major cause of the downfall of the Spanish at the turn of the twentieth century and today the actions of government institutions continue to alienate public opinion. The failure of the present day civil service to eliminate the failings of the past regrettably stems from the continued presence of the basic causal factors—inexperienced, indolent, corrupt or status seeking individuals in key positions.

In assessing contributions made during the Spanish era to the present day geography of the Philippines, note must also be made of the contribution of the two other exotic influences—Islam and the Chinese. Moslem south and south western Philippines languished as a cultural backwater for, as noted above, Spanish resistance turned the area into a remote and neglected end point of Islam cultural diffusion in South-East Asia. In contrast the focal point of Chinese influence, Manila, flourished, for the Spanish and upper class Filipinos tolerated the Chinese presence. Apart from international trade in the form of the Manila Galleons, the upper classes attached little social status to commerce. Furthermore, the passive attitude of the Chinese to political matters was favourably regarded by the administration notwithstanding periodic oppression, which had racial overtones. The adaptable Chinese filled the vacuum between rich and poor and speedily came to control most aspects of wholesale and retail trade, particularly in the northern half of the country. They also developed secondary industry and initiated market gardening. By their efficiency, persistence and adaptability they continue to exert strong influence on these areas of the economy but their influence has penetrated further afield both by racial mixing and by cultural innovation in the form of tools, handicrafts, crops and cuisine.

The American Era

The twentieth century ushered in for the Philippines a new era, for the country found itself with another Occidental master—the United States of America. The Americans, by and large, were reluctant colonisers and one of the keystones of their policy was the preparation of the country for independence as quickly as possible. In so doing they hoped to profit from the experience of other colonial powers, from their own colonial past, and from the failures of the Spanish regime they had succeeded. In many fields therefore the Americans were able to exercise their authority with sympathy and understanding.

Unfortunately, the American presence also introduced developments which were not only to negate progress made in certain areas but, more important, were to mould significant sectors of the Filipino way of life and livelihood in a retrograde manner. Thus, there was a tendency to formulate overambitious programmes whose objectives although well intentioned proved to be misguided, the emphasis upon liberal arts in education being a case in point. In addition, it proved impossible to introduce the beneficial aspects of American culture such as political democracy without including its less desirable attributes such as graft and corruption. On the Filipino's part, there was in general an uncritical acceptance of American influence. At the turn of the century, of course, any alternative to the increasingly resented Spanish rule was welcome but a more significant factor was the Filipinos' traditional lack of initiative. This was to become accentuated in the American era for not only were the Filipinos treated by the Americans as 'little brown brothers' but the physical stature of most Americans, coupled with their obvious socio-economic superiority, ingrained in many Filipinos a pronounced inferiority complex.

In the political arena the successes and failures of the Americans are well marked. The concepts and practices of democracy were speedily disseminated and readily absorbed; in like manner, a constitution was drafted along the lines of its American counterpart, and legal and judicial institutions were established. But, despite the erection of this framework, it was found difficult to make it viable. Thus, universal franchise led to ballot rigging by unscrupulous persons of wealth and influence and judicial impartiality in many areas was thwarted by similar improper pressure. For many Filipinos, therefore, the iniquities of the Spanish era persisted. Needless to say, strenuous and often successful efforts were, and are, made by the authorities to stamp out these practices. Complete success, however, cannot be attained until behaviour patterns are no longer conditioned by traditional modes of conduct. Such modes, ingrained for over three centuries yet disapproved of for only half a century will take time to disappear. All the more so as they apparently are condoned by influential elements in Filipino society, and in America, the homeland of a culture which is regarded by many Filipinos with a respect bordering on awe.

Paralleling the developments in politics were the Americanisation and broadening of the administrative framework. Again, Filipino inexperience and temperament did not, and still does not, permit its efficient operation. Excessive red tape and other failures of the system frequently hampered the implementation of government policy, itself already delayed and distorted by political infighting on the part of vested interests. Efforts to develop natural resources are a case in point: finance for ambitious survey and inventory programmes for pioneer areas was often drained away into established agricultural areas with greater political pull. Thus, many natural resources remained unknown or untapped, notably the country's mineral wealth, or else were exploited without regard to proper conservation measures, logging being a case in point. Greatest progress was therefore made in those areas which lacked vested interests and which were beyond the scope of private enterprise. Education and public health stand out in this respect, although even here a lack of impartiality and inefficiency on the part of the administrators resulted in uneven progress.

The Americans were also successful in creating a secular administration, although the Church retained an indirect influence. The formal break with the Church nonetheless made possible partial integration of non-Christian minorities and as a consequence, both headhunting and Moro raids largely died out.

Education was a particularly potent integrating factor. Free and universal educational facilities were provided as fast as possible by the authorities. The school house became the symbol of American colonisation and like the comparable Spanish symbol—the Church—was staffed initially by nationals of the colonial power. The influence of the American schoolmaster was equally as potent as that of the Spanish priest and he spearheaded the dissemination of American culture, particularly its more beneficial aspects. It was the language of instruction—English—which proved to be the integrating medium. Unlike Spanish, English came to be widely understood and also spoken, albeit imperfectly at times, by most levels of society. More so than the official National Language (Pilipino) it is the *lingua franca* of the islands.

Other actions by the Americans helped to make the nation more closely knit. Transport and communications underwent a transformation, although it must be admitted that principal activity in this sphere was concentrated in the more populous areas. Nonetheless, movement within the archipelago was greatly facilitated, which not only made possible the dissemination of American cultural influences but also accelerated the two opposed population movements: to the metropolis of Manila and to the pioneer frontier of Mindanao. External links too were expanding: in the Spanish era prior to the mid-nineteenth century regular contact had been established only with China and Mexico but now the steamship and later the aeroplane made all parts of the world accessible. Despite this, the overwhelming bulk of intercourse was with the United States for, in order to speed the development of the Filipino economy and in response to pressure by United States business interests, a system of free trade was initiated. This made possible a substantial expansion of agricultural production in the Philippines, which in turn created a sizeable market for American industrial products. But the peripheral effects were to prove more fundamental. They included privileged access to sources of developmental capital and technological skill which dwarfed that provided by the Chinese in the Spanish era. In large measure it was agriculture that benefited, especially the processing of agricultural products, but sufficient 'fall out' materialised to encourage other means of livelihood, e.g., vehicle body building. More effective use, however, could have been made of the situation had not a combination of Filipino temperament and American educational bias towards the liberal arts elevated the status of white collar jobs, and in particular those of least value to a developing nation—the law and politics.

The presence of the Americans also made the country a much more healthy place. The reduction in violence on the part of the headhunters and the Moros has already been commented upon but this was unfortunately offset by an increase in violence in urban areas largely as a consequence of the adoption by the naturally emotional Filipinos of some of the less desirable aspects of American political life which subsequently extended alarmingly into criminal activity. On the public health front, however, great strides were made. Epidemics were brought under control spearheaded by a degree of malarial control paralleled in South-East Asia only by Malaysia. In turn, the falling death rate caused a dramatic rise in population, for the retention of Catholicism by the population caused the birth rate to remain at a high level.

The problems created by the Americans in respect of the population explosion resulting from their public health measures is but one of several ways in which American intervention caused new problems to replace old ones. The abuses of democratic procedures have already been commented upon, as have been the problems created by a liberal arts education. In like manner road improvements led to ribbon development, particularly in the vicinity of Manila (paralleling this, security in rural areas reduced the significance of the nucleated settlements characteristic of the Spanish era). Lack of planning was also evident in respect of population growth in rural areas, where official settlement schemes proved to be failures, and in those booming urban centres where the creation of the Chartered City often merely offered greater scope for graft and corruption by local politicians. In certain sectors, too, leniency on the part of the Americans led in a very short time to an influx of cultural influences too massive to be assimilated by the Filipinos. In particular, there were those beamed at the lower levels of society who in the Spanish era had been insulated by ignorance and poor communications. American-style advertising coupled with Chinese exploitation of the economic potential broke down this insulation and exposed the common man to alien ideas and habits such as the power of money, comics, soft drinks and cosmetics. Such developments not only provided undesirable distractions for a developing nation but, more important, also created a demand which has grown so spectacularly as to require a significant proportion of scarce industrial capital resources. However, it should not be forgotten that by the same process—advertising and the Chinese businessman—beneficial influences such as toothpaste, technical manuals, and patent medicines were simultaneously introduced. In general, the influx of American-style consumer products has been to the advantage of the ambitious Filipino by stimulating his mind and/or inducing him to work harder to satisfy his new desires. On the other hand, they have further debased the lowest levels of society, particularly that which is urban based. And, of course, they have had an adverse effect on the balance of payments, the most striking example of which is, perhaps, the exceptionally high level of wheat and wheat flour imports to satisfy the Filipino taste for bread.

Many rural inhabitants, too, have lost ground as a consequence of the American era. After the defeat of the Spanish it was confidently expected that one of the root causes of discontent—exploitation of the peasantry by the *principalia* (Spanish and Filipino landlords)—would be eradicated. But this proved too complex a problem even for the Americans to solve despite a succession of land reform measures. These foundered either in the political arena through opposition by vested interests or else by their complexity attracted too great a burden of red tape. Only in the case of the friar lands was the agrarian situation alleviated and this merely whetted the appetite of the peasantry for further progress. Subsequent frustration culminated in armed revolt by the Hukbalahaps. But, with the landowners' position entrenched by the Americans' formal recognition of their tenure rights, even this development failed to produce a solution.

The American era lasted but a short time. Yet in it far more was accomplished than throughout the Spanish era. The Occidental element in the Filipino way of life and livelihood was markedly boosted, methods of livelihood and living standards were modernised, and the country gained its freedom. Yet the very pace of change created almost as many problems as it solved. Some of these had their roots in the Spanish era, others were new creations. In like manner the Americans both improved facets of the Spanish era and introduced alien concepts. Few of these developments, and especially the innovations, have been fully integrated into the Filipino way of life and livelihood. Unlike other emergent nations, therefore, the Philippines is in the process of both adjusting to the consequences of its independence and to the consequences of colonisation. When this is taken into account, and also the adverse effects of occupation by the Japanese in the 1940s[6], it is remarkable that the current state of flux in the geography of the Philippines is not far greater than it is.

References

1. T.S. Oracion, 'Kaingin Agriculture Among the Bukidnons of South-Eastern Negros, Philippines', *Journal of Tropical Geography*, Vol. 17, 1963, pp. 213-24.
2. F.M. Keesing, 'The Isneg: Shifting Cultivators of the Northern Philippines', *South-Western Journal of Anthropology*, Vol. 18, 1962, pp. 1-19.
3. In terms of religion the *barrio* was the site for chapels visited occasionally by the *poblacion* priest. He enticed his flock to regular worship in the *poblacion* church by means of colourful ceremonies incorporating or imitating traditional pagan initials, ceremonies which were climaxed by those associated with feast days of local saints (the *fiesta*) and other important dates in the Catholic calendar.
4. Though supplemented by the Chartered City. This American innovation removed selected major urban centres from the control of the provincial government and placed them within the framework of the presidential administration. This laudable effort to tackle the special problems associated with urbanisation has been only partially successful as resultant opportunities for political patronage and power have often been exploited. Manila (1910) and Baguio (1909) were the first chartered cities: increasing numbers have been added since 1935 and currently the total exceeds forty.
5. C. Vandermeer, 'Population Patterns on the Island of Cebu, The Philippines: 1500 to 1900', *Annals of the Association of American Geographers*, Vol. 57, 1967, pp. 315-37.
6. A United States Congressional Report in 1945 stated:
'Official reports, photographic evidence and statements of those who have seen the ruin and destruction are unanimous in asserting that, of all the war-ravaged areas of the world, the Philippines are the most utterly devastated from the standpoint of the ratio of functional construction still intact to functional construction damaged and destroyed, the effect of destruction on functional economy, social facilities of the nation, and the effect of the war damage on the capacity of the nation to rebuild and repair.'

8
The Present

The Philippines has been independent only since the end of the Second World War. The country, therefore, still is digesting the fruits of half a century of American control, whose impact in terms of pace and penetration far outstripped that of the Spanish. As was the case with those introduced by the Spanish, American institutions are being adjusted to conform to Filipino values but for the moment a state of flux exists.

The Philippines is one of the world's developing nations, and as such its geographical patterns are characteristically impermanent. Any comprehensive review of the distinctive geographical characteristics of the country today is supported best by investigations of specific features of the life and livelihood of the nation. Such case studies need to illustrate the national and regional implications of the dynamic processes currently at work, processes whose origins have been analysed in the preceding chapter and whose future is predicted in the subsequent one. Therefore, present day economic and social geography being the theme of this chapter, a general survey of the Philippine economy at the end of the 1960s is reinforced by discussion of selected sectors crucial to the Filipino livelihood. Similarly, the review of current social conditions contains studies of some of those issues likely to have a significant impact upon the Filipino way of life.

A. Economic Geography

As befits a developing nation, the Philippine economy is primarily agricultural in nature. Manufacturing, however, makes a substantial contribution: in 1969 it contributed 17·1% to total Philippine NDP of £1750 million at constant (1955) prices

TABLE 24
Net Domestic Product (NDP) at constant (1955) prices

	1963 £m	1963 %	1967 £m	1967 %	1968 £m	1968 %	1969 £m	1969 %	% Growth 1963-9	1970[a] %
Agriculture, Forestry and Fishing	408.7	33.4	484.1	30.7	504.7	31.4	552.0	31.5	35.1	33.2
Mining	18.5	1.5	27.3	1.7	30.3	1.9	37.7	2.2	103.8	2.0
Manufacturing	220.5	18.0	273.5	17.4	278.0	17.3	299.9	17.1	36.0	19.1
Construction	45.4	3.7	59.4	3.8	60.1	3.7	53.2	3.0	17.2	2.5
Transport, Communications and Utilities	58.4	4.8	73.9	4.7	74.1	4.6	80.6	4.6	38.0	3.9
Trade, Banking, Insurance and Real Estate	164.6	13.5	235.7	15.0	236.3	14.7	254.7	14.6	54.7	16.0
Government and Other Services	307.4	25.1	421.0	26.7	426.0	26.4	472.3	27.0	53.6	23.3
Total	1223.5	100.0	1574.9	100.0	1609.5	100.0	1750.4	100.0	43.1	100.0

Source: National Economic Council.

[a] Absolute figures on which these are based are at constant (1967) prices and thus not comparable.

(Table 24). Agriculture (including forestry and fishery) contributed a more substantial 31·5% but the sector has grown by only 35·1% in the period 1963-9, significantly below the overall growth rate of 43·1%. Manufacturing (36·0%) was equally sluggish. It was tertiary activity (excluding transport) which contributed most to NDP growth although the principal growth sector was mining which recorded a 103·8% increase in the period 1963-9.

In the period 1963-9 Philippine GNP rose by 37·8%. However, whereas in both 1968 and 1969 a 6·3% rate of income was recorded, the 1970 rate fell below 5% as a consequence of the stabilisation programme which followed the devaluation of the *peso*. Real GNP growth recorded in the 1960s, however, is not considered adequate for a developing nation such as the Philippines, and compares unfavourably with growth in earlier years of independence when it was around 7·5% per annum. The slowdown in GNP growth can be attributed to the various fiscal measures resorted to by the government to keep the nation's trade deficit within bounds. In fact, the real gain in the 1960s was barely 2%, for the country's population is growing at an annual rate in excess of 3%.

The Philippines, like other developing nations, is heavily dependent upon world commodity markets. The special trading relationship with the United States provides a valuable buffer but fluctuations in export earnings of basic primary products have an immediate and substantial impact upon the economy. A case in point occurred in 1968 when reductions in output of two major export earners—coconut and sugar—resulted in strong pressure on the *peso*. In recent years the *peso* consistently has been relatively weak for the country has been running a substantial deficit on its balance of payments current account. Since 1964, United States expenditure directly attributable to the Vietnam War has bolstered up the economy (such spending amounting to up to 20% of GNP) but has also contributed to monetary inflation by generating strong import demand. Periodic governmental measures to stabilise the economy have proved ineffective with the result that the 1960s proved to be a period of relative stagnation for the economy.

The work force of the Philippines is characterised by a high degree of unemployment (7-8%) with an even higher incidence of under-employment (around 25%). The total labour force in

1970 was estimated by the National Economic Council to be in the order of 13·2 million persons, two-thirds of which were located in rural areas (in contrast, the second largest employment sector—manufacturing—claimed barely one tenth). Organised labour is in its infancy and this, coupled with the over-supply of labour, makes for low wage levels. This state of affairs applies both to unskilled labour and to the substantial proportion of relatively well educated workers (45% of the labour force have had at least six years of education and 15% have had high school training or better). Wages remain low despite a substantial increase in 1970 in the minimum daily wage to £0·85 for industrial workers and £0·50 for agricultural workers.

Maldistribution of wealth is an outstanding feature of the Philippines and has social as well as economic implications. In 1970 it was estimated by the Social Communications Center that the middle and 'rich' classes of Philippine society (those earning more than £200 per annum) comprised only 20% of the population; in contrast, of the less privileged members of society one in three received less than £100 annually. The bulk of the nation's wealth, whether it be land, industry, commerce, or political power, lies firmly in the hands of those described by the Center as the 'super rich'; such members of society, however, comprise an infinitesimal 0·07% of the population. Regional disparities also are striking: the median family income for the fiscal year 1967/68 was estimated at £175 (or about one eighth of that for the United Kingdom) but in Manila comparable incomes were nearly double national levels. High income families, i.e., those earning £500 or more, in 1968 accounted for one in four of all families surveyed by the Bureau of Census *Survey of Households*, but nationally such families are confined to less than one in ten. Over two out of every three of the nation's families earned less than £250 in 1968,[1] whereas the comparable figure for Manila was two out of every nine.

Primary Production

Agriculture is far more important than any other aspect of primary production despite important contributions from mining, forestry and fishing. Agriculture provides over half the total employment opportunities, indirectly supports about two-thirds of the population, and accounts for over three-quarters of total exports.

Low productivity in the food crop sector is characteristic of all but minor areas of the country. It is caused by a variety of factors notably the small size of farms, inadequate irrigation, inefficient cultivation techniques, distribution bottlenecks and farm tenancy. As a result, yields of the staple products—rice and maize—are amongst the lowest in the world and in the past have necessitated heavy imports with subsequent strains for the economy. The successful introduction of IR-8 and other high yielding varieties (HYV) of rice has modified remarkably quickly the rice situation.[2] In 1970, planting of HYV rice occurred on about 50.3% of the nation's rice lands but a continuation of the dramatic expansion of recent years appears likely to be frustrated by the lack of capital available to tenant farmers and others equally willing but financially unable to utilise the new seed successfully. Possible difficulties resulting from increased production will thus be postponed, forcing down prices and squeezing out marginal (and therefore small) producers.

Yet, surprisingly, it is the cash crop sector which gives greater cause for concern. In the 1960s value added by food crops rose by about 11% per annum at current prices, whereas the comparable figure for cash crops was only 8%. Only coconuts of the major cash crops recorded a substantial increase in production value and this was more than offset by a decline in the value of tobacco, abacá, other fibres and rubber. Factors responsible for this unimpressive performance include those already mentioned for food crops reinforced by recurring natural disasters, notably typhoons, constitutional barriers to large scale production, and increased competition from other tropical producers (principally faster developing nations, such as Malaysia) and from synthetic substitutes.

Typhoons not only depress national agriculture output but have a disastrous impact upon agriculture in those localities that bear their full brunt. Thus, in 1968 Negros Occidental sustained losses totalling £17·5 million from typhoon Senyang, the worst ever to hit the area in living memory. The typhoon rendered 44 554 people homeless and killed 122. It also destroyed 10 687 houses, 2226 classrooms, twenty-eight bridges, three sugar *centrals*, 315 livestock farms, and ten fishing boats. Total damage to sugar, rice, maize and other agricultural crops amounted to £1·5 million. At least 100 000 tons of sugar were lost due to the resultant floods and the stoppage of milling operations by the destroyed *centrals*. More recently, a series of typhoons were experienced in late 1970 which destroyed hopes of a 10% increase in coconut production for 1971. The province of Catanduanes, the most vulnerable part of the typhoon-prone Bicol Peninsula, suffered the destruction of up to 90% of its coconut trees.

Less spectacular than typhoons, but in the long run more damaging, is soil erosion. In their natural state most Philippines soils are deficient in plant nutrients with the result that the application of artificial fertiliser is necessary for efficient agriculture almost without exception. Until 1952, all fertiliser employed (only a relatively small amount—40 000 tons per annum—because of the ignorance or lack of finance on the part of local farmers) was imported. The development of the hydroelectric potential of the Agus River at Maria Christina Falls made possible the establishment of a 50 000 ton ammonium sulphate fertiliser plant using local copper and iron pyrites. Subsequently, superphosphates have been produced at Toledo (Cebu) and Manila. The former, and larger, uses sulphuric acid from the local copper processing plant and imported phosphate rock. Despite these and other planned developments, and demand generated by the new HYV rices, it will be a long time before the application of artificial fertiliser reaches acceptable levels. This is self-evident from the fact that in 1967, 130 379 metric tons were consumed which is only the equivalent of a mere 15·3 kg per hectare of cropland (Table 25).

The magnitude of the task facing soil conservationists is further highlighted by the results of land capability assessments.[3] Of land considered suitable for cropping (44% of the total) only 27% is suitable for intensive cultivation. On the other hand, 47% can be successfully cultivated only if sophisticated soil conservation techniques are employed.

Development in the agricultural sector is being conducted on a number of fronts. Some programmes such as forest clearance in Mindanao make an immediate impact on the landscape,

TABLE 25
Fertiliser Consumption 1967

Crop	Crop Area ('000 hectares)	Fertiliser Consumption (metric tons)	Fertiliser Consumption (nutrient content) (kg per hectare)	Percentage Comparison Crop Area	Percentage Comparison Fert. Consumption
Sugar	309	67 131	217.3	3.6	51.5
Rice	3096	34 429	11.1	36.4	26.4
Maize	2158	4163	1.9	25.3	3.2
Tobacco	82	2495	30.4	1.0	1.9
Coconut	1820	2675	1.5	21.4	2.0
Other Crops[a]	1048	19 486	18.6	12.3	15.0
Total	8513	130 379[b]	15.3	100.0	100.0

Source: Economist Intelligence Unit.

[a] Principally vegetables, citrus, pineapple.
[b] In terms of total volume this amount represents some 500 000 tons; by 1969 this had increased by about 30%.

others such as the improved distribution of pesticides have a more subtle effect. But all of the measures currently in operation are designed to increase man's control over his environment and as such have geographical implications. Man, unfortunately, does not always appreciate the benefits to be obtained from innovations, particularly if he is an ill-educated *tao*, for the benefits of agricultural extension are slow to materialise, requiring both prolonged indoctrination and the means to finance change.

Agricultural extension is a vital element in overcoming ignorance and conservatism. But, it should not be forgotten that a lack of response is not solely the fault of the farmer. Farm mechanisation is a case in point where the advantage to the farmer is expressed in terms incomprehensible to the average farmer and which require considerable expertise on the part of the extension worker to produce both a cogent and coherent explanation. Thus, for example, the cost of a work *carabao* has been estimated at £57 while a six horsepower gasoline-fed hand tractor is estimated to cost from £65 to £83 per horsepower—but it is claimed that as regards horsepower, the slight difference in cost between the *carabao* and the tractor may be offset by the fact that the tractor is less risky to keep (the *carabao's* susceptibility to diseases has made it uninsurable). Furthermore, the use of a tractor on a one hectare rice field can reduce costs by £4 and an additional reduction of £1 can be effected via combined *carabao*-tractor power—in the latter case, while it is cheaper by £1 to use the combined power, the amount as opportunity saving in four days' time is smaller compared to the opportunity to earn in four days; in other words, while a farmer who uses the combined power saves £1

TABLE 26
Traditional and IR-8 Rice: Comparative Costs (£ per ha.)

	Traditional	IR-8	% gain by using IR-8
Production Cost	200	800	300
Yield (cavans)	30	100	233
Income (at £17 per cavan)	510	1700	233
Profit	310	900	190

Source: *Journal of Commerce*, Oct. 7, 1968.

by waiting four more days to have his field prepared, the farmer who employs a tractor has four days saved within which he is free to work and earn more than £1. Contrast this involved argument with the obvious merits of IR-8 rice (Table 26). Fortunately, the merits of mechanisation are getting through; in the past 70% of tractors have been purchased by sugar and large scale rice producers but now 70% are being bought by small rice producers on a co-operative basis.

In the food crop sector self-sufficiency is the major development goal; this involves increasing the area under irrigation, farm mechanisation, expanding processing facilities, wider use of fertilisers and pesticides, the provision of animal feeding stuffs, improved varieties of seed, and land reform. Such developments may also have major implications for the cash crop sector but here there is also a need for more sophisticated marketing procedures—diversification of markets, improved product quality, and a sustained promotional effort—together with investigations of alternative uses, the encouragement of local manufacturing facilities, and replanting incentives. Some of the results of these developments are illustrated below, using a food product (meat) and a cash crop (abacá) as examples.

Meat. With meat, as with certain other food items, notably fish, dairy products and wheat flour, Philippines consumption exceeds any domestic production with the result that scarce foreign exchange is needed to finance imports. Livestock production in the Philippines has been handicapped by climatic conditions but these now can be overcome with the use of modern techniques. However, the fragmented and inefficient nature of production restricts the application of such techniques. Even where they are employed, producers run up against other problems, notably inadequate irrigation facilities and distribution bottlenecks.

In recent years both livestock numbers and domestic meat production have been trending upwards (Table 27). Such trends are particularly noticeable for pork, which accounts for more than four-fifths of all red meat eaten in the country. Large scale production has been an especially potent factor; in the past decade average herd sizes have more than doubled, to 1110 for pigs and 400 for beef cattle. One piggery near Manila contains 19 000 animals. There is also an 800 000 unit poultry farm.

TABLE 27
Livestock Statistics, 1965-70

	1965	1966	1967	1968	1969	1970
Annual Population ('000)[a]						
Carabao	3346	3633	3926	4173	4369	4432
Cattle	1560	1583	1579	1644	1629	1679
Pigs	6939	6914	5497	6090	6350	6456
Goats	606	616	599	624	698	772
Chickens ('000 000)	57	68	66	68	63	57
Red Meat Production ('000 metric tons)						
Beef/Carabao meat	85.0 (est.)	92.5	80.9	75.3	460.0 (est.)	450.0
Pork	290.0	317.0	340.0	366.9		

Source: Bureau of Agricultural Economics and United States Department of Agriculture.

[a] On farm population only, the total population is much greater, e.g., cattle: 2 million, pigs: 12 million, goats: 0.9 million.

Carcase weights have shown considerable improvement also but the discrepancy between commercial and peasant, or semi-commercial, operators (180 kg to 110 kg in the case of cattle) remains wide.

Nevertheless, at present two-thirds of all meat consumed has to be imported. The root cause of this situation is that increased production is not keeping pace with increased demand which in turn results from the country's rapid population growth as much as from increases in per caput consumption. The latter remains low: about 14 kg per annum, or about one sixth of that for the United Kingdom. Furthermore, per capita consumption is likely to continue at a low level as the current stagnation in growth of real income per caput is expected to inhibit purchases of red meat which to many is a luxury item.

The upward trend in livestock production has been achieved despite the presence of several obstacles to growth. A basic problem is the scarcity and cost of quality animal feedingstuffs. Maize is available but it is expensive and local feed millers often reduce the quality of their feed to offset rising maize prices. Protein supplements are unavailable in sufficient quantity locally and imports are prohibitively priced because of the high duty. On the land itself the inadequacy of irrigation facilities means that in areas with a pronounced dry season, such as the

Central Plain of Luzon, forage is insufficient, while in areas such as the Bukidnon plateau of Central Mindanao, where climatic conditions permit natural all year round growth, forage quality is usually poor. Efforts by the Bureau of Animal Industry to encourage livestock development are hampered by a lack of funds and of trained personnel; for example, only a handful of better breeding stock have been delivered to farms and ranches to upgrade local cattle under the provisions of the Animal Dispersal Act. Lack of trained personnel is also a major problem on production units; qualified animal husbandmen are rare, especially those able to handle large scale operations.

At present, growth in the livestock industry is in large measure haphazard and is stimulated primarily by the high price meat can command. Furthermore, the bulk of meat produced is consumed on the farm or is destined for sale on adjacent markets to meet demand generated by *fiestas* and other celebrations which, for many, represent the only occasions when red meat is consumed in any quantity. Such semi-commercial production is a universal element of the rural scene and, in the case of pigs, the urban scene as well where the smaller centres are concerned.

True commercial livestock production is of significance only in parts of Luzon, especially Bicol, and Mindanao (Fig. 35). Even in such areas operations designed to produce high quality meat are the exception. The problem here is the difficulty of establishing efficient finishing procedures due principally to the limitations imposed by the supply of animal feed mentioned above. Pig producers located in or near urban areas have access to food waste but beef producers can only produce fat cattle if they are the few fortunate enough to possess their own feedmixing facilities or, like those in Bukidnon, have access to a good source of cheap raw material in the form of pineapple waste.

Abacá. The Philippines has a virtual world monopoly of abacá production but in recent years the industry has been far from prosperous. The roots of the industry's problems go back to the Second World War, but subsequent events have multiplied the difficulties facing abacá planters. In brief, abacá production today suffers from most of the problems facing cash crop production in the Philippines and from typhoons, synthetic

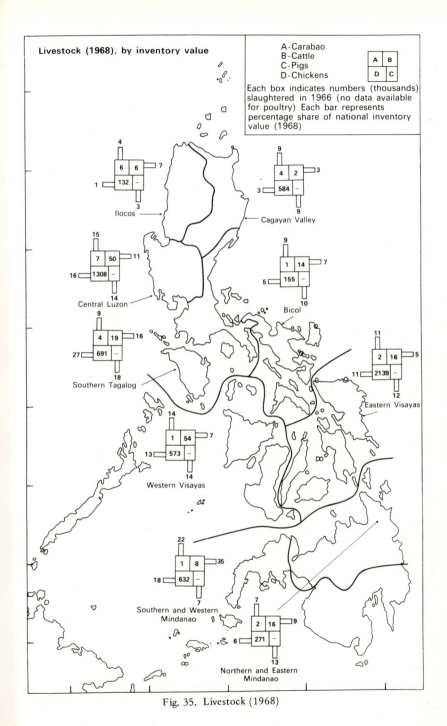

Fig. 35. Livestock (1968)

substitutes and inefficient production techniques in particular. As a result, abacá is becoming a less prominent feature of the rural scene as producers switch to more remunerative crops.

The problems facing the abacá industry have been realised for some time; in 1955 the Abaca Corporation of the Philippines (ABACORP) was set up to supervise and promote the development of the industry. Unfortunately, like many such bodies in the Philippines, its activities when funds were available (which frequently they were not) did not match its ideals. ABACORP in fact became preoccupied with the export of abacá and the domestic problems of the industry were neglected. Therefore, to overcome ABACORP's deficiencies, its successor—the Abacá and Other Fibres Development Board—is designed specifically to provide agricultural assistance and to search for new outlets and new uses for abacá.

Typical of past misplaced effort was the government's appropriation of over £2 million for the development of the abacá industry. It was planned to increase annual output by 25 000 to 30 000 metric tons by bringing back into production 49 000 hectares, equivalent to more than one quarter of the existing area under production. To encourage this shift in production, Development Bank loans to planters were increased from £55-85 to £105 per hectare. While laudable in its intentions, this development took little account of the fact that the basic problem is one of production exceeding demand. The fact of the matter is that overseas demand for abacá continues to be adversely affected by synthetic substitutes, especially in the field of its principal end use—cordage.

Philippines producers' ability to compete in overseas markets has been made that much more difficult by local conditions. For example, typhoons in 1970 were reported to have destroyed over one fifth of abacá plantings, with damage concentrated in Bicol where the industry was looking forward to a brighter future following the employment of abacá as a paper pulp material. The vulnerability of the industry to earlier severe typhoons, notably in 1964 and 1967, which badly damaged existing plantations in the Visayas and Mindanao, and the lack of progress in checking the ravages of disease have been a great discouragement to producers. It has been commonplace for production to be abandoned or else intercropping of coconut substituted. With low prices forcing down wage rates below the legal minimum a redeployment of the labour force is also taking place. The numbers involved are not known but a significant proportion of the 3·5 million Filipinos who formerly were wholly or partially dependent on abacá for their income must no longer have any connection with the product.

Manufacturing

Large scale manufacturing in the Philippines had only just commenced under the stimulus of government owned corporations formed at the beginning of the Commonwealth period, when the Second World War intervened. War damage funds were only partially used to rehabilitate manufacturing facilities, being diverted into consumer goods and commerce. It took the resultant economic crises to stimulate an awareness of the benefits of industrialisation. Now government economic policy is committed to 'industrialisation, particularly in the development of a strong manufacturing sector. It is committed to the development and maintenance of private entrepreneurship, although government funds are frequently employed in the creation of large government corporations that operate in what is normally considered a private sector of the economy. The formation of these corporations encourages the flow of private capital into those sectors of manufacturing in which it is reluctant to enter'.[4]

Industrial growth remains centred upon processing and assembly operations (Table 28). Thus, production is concentrated upon food, beverages, tobacco products, textiles, clothing, rubber products, pharmaceuticals, paints, plywood and veneer, small appliances, and automobiles. Heavy manufacturing contributed less than 40% of total value added by the manufacturing sector in 1969 with cement, glass, industrial chemicals, fertilisers, iron and steel, and refined petroleum the principal products involved. The 1960 Census revealed that over two-thirds of manufacturing employment was to be found within 100 kilometres of Manila (Fig. 36). Beyond, manufacturing is very much conditioned by raw material sources, notably sugar (Negros Occidental's sugar industry accounted for about 8% of the nation's 1960 manufacturing employment or about one quarter of that located more than 100 km from Manila). In

TABLE 28

Manufacturing Employment in the Philippines, 1960 and 1969

Major Product Group	1960[a]			1969[b]	
	Greater Manila	Provinces	Total	No. of Establishments	No. of Employees
Food and Kindred Products	12 587	45 397	57 984	2923	82 956
Beverages	5107	3270	8377	94	15 027
Tobacco Products	9172	1426	10 598	57	20 828
Textiles	15 654	15 703	31 357	211	47 319
Footwear, Apparel and Related Products	15 578	16 395	31 973	2826	39 083
Lumber Products	5381	14 924	20 305	531	38 512
Furniture and Fixtures	2860	1223	4083	414	8288
Paper and Allied Products	3439	692	4131	96	7880
Printing and Publishing	11 016	1692	12 708	518	15 344
Leather, Leather and Fur Products	309	436	745	65	2517
Rubber Products	5614	600	6214	105	8450
Chemical and Allied Products	8546	2042	10 588	335	24 046
Petroleum and Coal Products	41	470	511	15	1886
Non-Metallic Products	4779	3491	8270	458	17 598
Basic Metal Products	2199	856	3055	61	9191
Fabricated Metal Products	9657	1431	11 088	477	23 139
Machinery (except electrical)	3064	234	3298	336	6107
Electrical Machinery Products	6184	281	6465	207	12 261
Transport Equipment	4870	2860	7730	503	15 393
Miscellaneous	3647	806	4453	282	7565
Total	129 704	114 229	243 933	10 514	403 390

Sources: (a) T. W. Luna Jnr, 'Manufacturing in Greater Manila', *Philippine Geographical Journal*, Vol. 8 nos. 3-4, July-Dec. 1964.
(b) Bureau of Census and Statistics: *Annual Survey of Manufacturers, 1969* (no regional breakdown available).

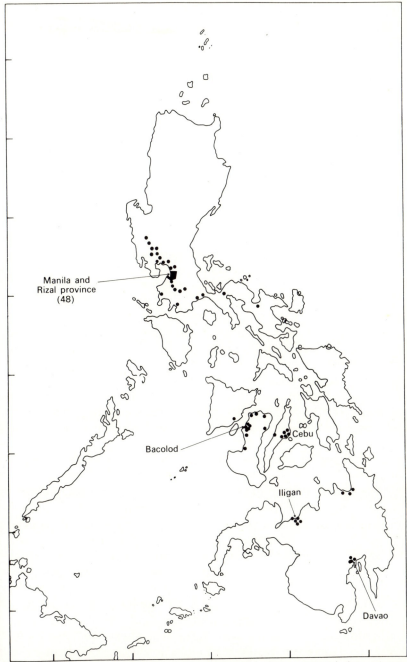

Fig. 36. Manufacturing (based on latest available—1960—provincial data; recent changes are believed to cause no substantial change to the pattern shown)

1960 manufacturing employment in Mindanao was only 11% of the nation's total, but the Misamis Coastal Belt, and Iligan City in particular, is emerging as an important focal point of manufacturing activity. When to this development are added the various primary product processing facilities being established on the island in recent years, it is not surprising to find that one third of all projects approved by the Board of Investments are now located in Mindanao.

Net value added by manufacturing increased by an average of 10% a year at current prices from 1962 to 1969 but only by 6% in constant (1955) prices. The sector contributed £300 million to NDP in 1969 (see Table 24); this represented 17·1% of total NDP, marginally less than its contribution in 1963 (18·0%). In 1969, 11·9% of the work force was employed in manufacturing; manufacturing employment overall increased by 61% between 1960 and 1969; large scale manufacturers account for about four-fifths of total output but employ only one quarter of the labour force engaged in manufacturing. The barriers to a faster rate of growth in manufacturing include a weak capital base, restrictions on overseas investment, heavy accounts receivable, shortage of credit and marginal profits. High production costs restrict the scope for exports and cause a reliance upon local demand which often is neither substantial nor stable. For example, attempts to increase the value added by timber manufacturing failed in respect of pre-finished plywood as, although there was a ready export for first class material, the local market could not absorb the balance of production of below export quality essential for a balanced profitable manufacturing level.

Efforts to strengthen the manufacturing sector include protective import duties and taxes, re-financing of distressed industries, and tax incentives both for preferred and pioneer industries and for exports. Early in 1971 the government relaxed controls on foreign investment to stimulate industrial development particularly in those sectors not proving attractive to local investors such as dairying, copper smelting and chemicals.

Priority manufacturing projects include integrated wood processing and pulp and paper production, industrial salt and soda ash, antibiotics, cement, agricultural machinery, machine tools, and cold storage facilities. Emphasis on such industries is designed to develop basic industries and to consolidate and expand existing facilities in this sphere. In addition, special attention is being paid to those industries producing short supply articles needed to increase agricultural productivity. For all priority projects a total of £180 million was allocated under the Four Year Plan which expired in 1970.

The result of this and other investment will be to cause significant changes in parts of the Philippines urban scene and also to accelerate the establishment of industry in rural areas. Thus, for example, developments in food processing, which accounted for no less than 24% of nett value added by manufacturing in 1969, are focused on the construction of several new sugar mills in rural areas (see Fig. 14). In contrast to these incipient development nodes is the case of textiles where the emphasis is on measures to ensure that existing urban facilities operate much closer to full capacity.

In order to illustrate more fully the problems and potential of development it has been thought desirable to provide detailed appraisals of a processing operation (copra crushing) and an assembly operation (motor vehicles). These are the two types of industry which, now and in the immediate future, form the nucleus of manufacturing in the Philippines.

Copra Crushing. Valued at £35 million in 1970, exports of copra represent one tenth of the country's exports and make the Philippines the world's leading supplier (it supplies about half of the world's copra needs). Exports of coconut products, notably coconut oil, are even more valuable totalling some £48·6 million in 1970 (see Table 7). Coconut oil contributed £40·5 million of this total and this product gives rise to one of the more important of the country's agricultural processing industries—copra crushing (since 1965 the export value of copra has been halved whereas that of coconut oil has increased by over 40%). Three-quarters of the coconut oil produced in the Philippines is exported: 255 649 tons in 1968, which was then the second highest total ever (1966 was then the peak year with 309 649 tons), rising to an estimated 320 000 tons in 1970, an all time record. In the United States coconut oil from the Philippines enjoys preferential treatment in the form of a substantial duty-free quota (80 000 tons in 1970) so it is not surprising that the United States absorbs four-fifths of Philippines coconut oil exports.

The combined annual capacities of the twenty-three existing copra crushing mills, of which twelve are substantial units, approached 1.5 million tons in mid 1972 (this represents an intake of more than 1·5 million tons of copra per day). Despite the considerable scope remaining to convert copra exports into more valuable coconut oil, the industry is operating at less than two-thirds of capacity, rising to three-quarters in the case of the largest producers who generally possess the more modern equipment. With total 1970 capital investment in the industry estimated at £30 million (in terms of replacement value), a considerable waste of resources is involved. It reflects in part the success of the industry in increasing the value of exports—between 1960 and 1971 copra's share of exports fell from 83 to 48% whereas the more valuable coconut oil rose from 10 to 47% (Fig. 37). In 1970, coconut oil became more valuable an export than copra for the first time (Table 7). Unfortunately, this success has attracted a disproportionate share of investment—a common phenomenon in the Philippines, resulting from the absence of controls and a belief by investors that in the long term continued growth (which does not always materialise) will eliminate over-capacity.

Most processing plant is located in the Manila metropolitan area. This is some distance from the main copra producing centres but the dominant locational factors would appear to be ease of exporting and access to local consumption. The Philippines imports sizeable quantities of oilseed products but these are imported already processed and thus do not pass through the Manila mills. Processing plant outside Manila is located mainly in Mindanao (Zamboanga City, Davao, and Jiminez to the north of Ozamiz City), but one of the most important producers is to be found in Cebu City.

Four more coconut oil mills came into operation in 1970 thus further inflating the over-capacity of the industry. The largest is located at Surigao, and produces edible oil for domestic use and also copra cake. The others are small units with capacities in the 100-250 tons per day range. Of the pre-1970 mills, 60% are foreign owned with both Unilever and Proctor and Gamble, for example, possessing substantial facilities in Manila. Despite the fact that the entire industry is operated by private enterprise the bulk of which is foreign based, government assistance has been provided to assist its

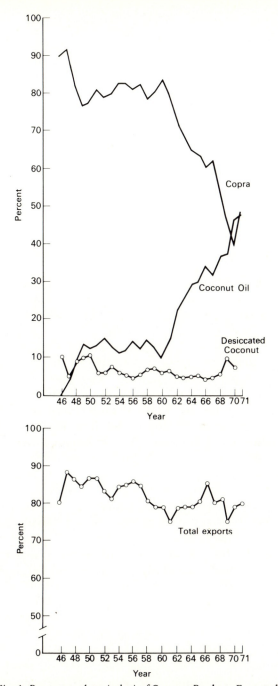

Fig. 37. Fig. A. Percentage share (value) of Coconut Products Exported, 1946-71
Fig. B. Exports as percentage of total value of coconuts produced

development for it is realised that the industry is of considerable importance to the nation. The government has, therefore, sponsored research into processing methods (as has private enterprise) via the Philippine Coconut Administration (PHILCOA) whose overall objective is to increase the competitiveness of Philippines products in world markets. In 1968 an integrated coconut industrial complex was inaugurated to produce plasticiser using coconut oil as the raw material. This was to be the first step in an ambitious programme designed to produce tannic acid and associated chemicals, and then soap detergents and margarine. Development Bank assistance, however, currently is focused on loans to planters for replanting purposes following the 1967 and 1970 typhoons which severely damaged certain of the principal production areas, notably Bicol.

The typhoon damage is expected to be the main short term factor affecting the oil processors. Until replanted trees (and, more important, substantial numbers of newly planted trees in Mindanao) bear fruit in the 1970s, copra production will be curtailed. Some advantage has been gained from the shortage of supply in the form of increased prices but any long term benefits accruing to the industry can only stem from the successful combatting of the three basic problems affecting the industry—the ending by 1974 of preferential trading with the United States, internal copra distribution bottlenecks, and rationalisation of existing capital investment, notably involving a control of new investment and channeling it closer to the copra producing areas especially those emerging in Mindanao.

Motor Vehicle Assembly. This ranks with such activities as oil refining and iron and steel as prestige industrial operations in the eyes of many developing nations. A desire to reduce expensive imports provides a complimentary attraction. The first motor vehicle assembly plant in the Philippines was established in Manila in 1951 but the benefits from this early entry into motor manufacturing have suffered from subsequent developments. A *laissez faire* attitude on the part of the government led to the emergence by the mid 1960s of over forty plants engaged in the assembly of about one hundred different models. This excessive fragmentation resulted in the average assembler gaining an annual market share of only 100-200 units per model. Subsequently, regrouping has

TABLE 29

Philippine Motor Vehicle Industry, 1965-70

	Passenger Cars	Registrations Trucks and Buses[a]	Total	Motor Vehicle Assembly Numbers Assembled		New Vehicle Sales		Total
				Cars	Trucks	Cars	Trucks	
1965	150 345	122 858	273 203	n.a.	n.a.	n.a.	n.a.	16 700
1966	174 394	134 743	309 137	n.a.	n.a.	n.a.	n.a.	18 500
1967	219 957	142 092	362 049	20[b]	25[b]	14 641	6986	21 627
1968	248 328	164 889	413 217	14	18	17 509	8515	26 024
1969	272 183	124 229	446 412	12	16	17 149	10 059	27 208
1970	279 172	179 445	458 617	11	15	7375	8824	16 199

Source: Philippine Automotive Association.

[a] No data available for tractors but annual sales are about 10 000 units, of which about 1500 are large units.
[b] In thousands.

occurred but twenty-six assemblers still were in existence in 1970 with total annual sales averaging 450-475 units only. In 1969, the Philippine motor vehicle assembly industry directly employed about 7000 persons, and total assets, including dealership operations, were in excess of £40 million. Northern Motors of Manila, in fact, is reputed to be one of the largest franchised General Motors dealers in the world. Ancillary industries also make a significant contribution to manufacturing activity in the Philippines. Of such industries, the most prominent is tyre production which is dominated by Goodyear, Goodrich and Firestone, whose fixed assets exceed £10 million.

The present irrational state of the industry is largely a result of import controls which, by restricting the supply of vehicles, assured the profitability of almost any motor vehicle distributor. In consequence, anyone who could establish a franchise fought for and gained a small market share. Such conditions are not uncommon in Philippines manufacturing but are hardly conducive to efficient operations on the part of producers, and hence realistic prices to consumers.

Nevertheless, the assemblers serve a rapidly growing market notwithstanding the deficiencies of the Republic's road network discussed below. In 1965 the number of vehicles registered in the Philippines passed the quarter of a million mark for the first time and by 1970 totalled 458 617 units (Table 29). At the end of the 1950s roughly equal numbers of passenger and commercial vehicles were in use but by 1970 the ratio was clearly in favour of the former, registrations comprising 61% of the total compared with 39% for commercial vehicles. As growth in passenger car sales was focused on Manila (Fig. 38), the assemblers, being located mostly in or adjacent to the metropolis, were ideally placed to serve the market.

In 1967 the assemblers' annual sales passed the 20 000 units mark for the first time. This represented a 17% growth on the previous year, but one of over 20% where passenger cars were concerned. The boom in the industry prompted President Marcos to dub it the nation's cultural barometer; however, low pressure was recorded in 1970 when sales fell to 16 199 units following the peak of 27 208 units set the previous year (1970 passenger car sales were only 7375 units compared with 17 149 units in 1969).

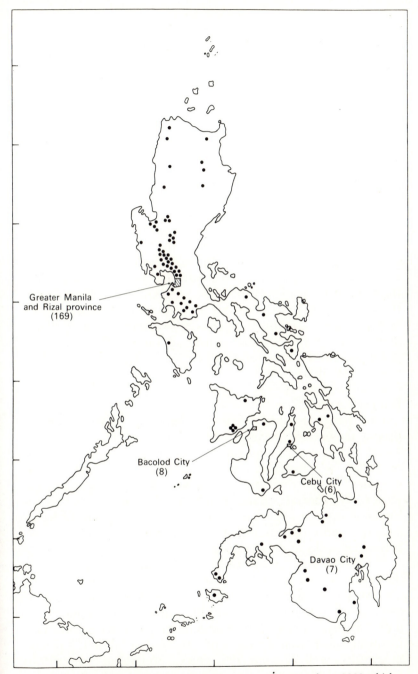

Fig. 38. Motor Vehicle Registration (Passenger Cars), 1970. One dot = 1000 vehicles

Transport and Communications

The pattern of economic geography in the Philippines is much influenced by the nature of its transport and communications systems. The physiography of the country has meant that in the past most movement has been waterborne. Now, increasingly, other modes of transport are assuming critical importance, and have been found wanting because of their poor condition and inefficiency. Despite the fact that its contribution to NDP in 1970 was only 3·8%, high priority has been given to investment in roads, bridges, harbours, etc.

In the Four Year Plan period up to 1970 over £50 million was spent on improving or extending roads totalling nearly 20 000 kilometres, or one third of the total system. Even so, the rapidly growing volume of road traffic and the all too frequent exceptional weather conditions mean that, apart from the Central Plain of Luzon, many areas still are badly served by the road network (for example, only 24% of the roads involved in First Plan expenditure possess all weather surfaces). This is particularly so in Mindanao where strenuous efforts are being made to remedy the situation (currently, the life of a vehicle tyre in Mindanao, and also in the Bicol Peninsula and the Cagayan Valley, is only a quarter that of one in the Central Plain of Luzon).

Rail transport, which is entirely government owned, does not play a major role; nearly 1000 kilometres of track exist on Luzon but it is the other rail system—less than 150 kilometres on Panay—which makes a more significant contribution to the economy by moving large quantities of both sugar and passengers. In contrast, poor quality of service means that the Luzon system in 1970 handled only 8·5 million tons of freight and little more than 8 million passengers, totals which are well below the peaks of previous years.

The insular character of the Republic causes it to possess over 350 ports some fifty of which are of national importance (Fig. 39). Manila is, of course, dominant, handling a net tonnage of cargo alone totalling over 25 million metric tons including the bulk of that entering international trade. Inter-island trade, involving substantial numbers of passengers (2 286 919 persons in 1969) as well as cargo (net tonnage 36 million tons in 1970) is conducted by means of some 200 ships of 100 net tonnage or

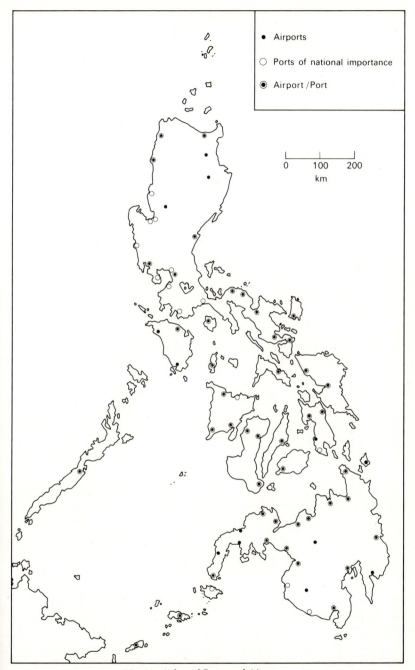

Fig. 39. Selected Ports and Airports

more (40% of which are based at Manila). The efficiency of inter-island boats suffers from the fact that, while they make over 1500 calls per week, the national ports have available less than 250 berths. The port of Cebu is preeminent in inter-island marine transportation handling 29% of the cargo moved in 1970 and 16% of 1969 passenger movements. Some 15 000 vessels call at Cebu annually.

Only 55-60% of the cargo handled by the inter-island network consists of commodities in the process of being moved from producers in one area of the Philippines to consumers in another.[5] The cargo outstanding is dominated, in volume terms, by the 6 million metric tons of commodities intended for eventual export[6] and, in value terms, by imported manufactured goods. From the point of view of national efficiency this local trans-shipment is undesirable. In part it is inevitable because of the fragmentation of centres of production and consumption, but it also reflects vested interests (notably the Port of Manila and the inter-island shipping companies) who profit from the existing situation and use the excuse of inadequate facilities to limit the establishment of new ports of entry (barely one third of the national ports currently are open to international shipping).

The fastest growing transport mode is air transport, handling over 2·5 million passengers in 1968. With domestic travellers totalling over 2 million, passenger movements soon will exceed those of the more traditional inter-island shipping services and, as such, traffic demands are running ahead of both airport facilities and aircraft. This is despite the fact that the country possesses over fifty airports which, because of the insular nature of the country and the mountainous island interiors, usually are sea-ports also (Fig. 39). To serve these airports, the three major internal airlines operate some 30 000 unduplicated route kilometres employing some 300 registered civil aircraft. More modern aircraft, especially those designed to operate from short and unsurfaced runways, now are more commonplace but domestic services still leave much to be desired. As such Philippine air transport represents one of several major barriers to the development of the archipelago's tourist potential; indeed, this and international standard accommodation can be regarded as a prerequisite if the overseas visitor is to be persuaded to extend his normal 1-2 days stop-over in

Manila to sample the attractions of Mt. Mayon, the Sulu Archipelago, Palawan, etc., at present accessible only to the adventurous.

Both government owned and private communications facilities extend throughout the country but the service provided cannot cope effectively with the demand. For example, the 1428 cities and municipalities in the country in fiscal year 1969/70 had access to only 1346 radio/telegraph/telephone stations. It was not until 1967 that the whole country was provided with a long distance telephone service, but by 1970 there was still only one telephone per 2300 persons. In contrast are the entertainment media: two out of three families possess a radio; there are 175 radio stations broadcasting in the Philippines; television channels now number fifteen; and colour television on three channels is available for viewers in Manila. Radio is a medium particularly well suited to the Philippines with its island environment, multi-lingual society and isolated rural areas; the case for investment in television, particularly in colour, is much less convincing.

The above is but a sketch of the nature and extent of transportation and communications systems in the Philippines. In order to demonstrate more clearly the implications of existing facilities for the economic geography of the country, it has been thought best to discuss in more detail a particular mode—road transport. This is the sector which received most attention (70% of available funds) in the First Plan and which also is receiving the lion's share of funds in the Second Four Year Plan commencing in 1971.

Road Transport. Highlighted by the inauguration of the 1300 kilometre long Pan-Philippine Highway designed ultimately to connect Luzon and Mindanao via Samar, road transport has come to play an increasingly important role in the economic geography of the Philippines. In the period 1967–70 the government undertook to increase the length of paved roads by some 40% (which still left some three-quarters of the network unpaved and only one tenth with a first class surface). At the same time, the total length of the road network was increased by 3000 kilometres, or by about 5%. Bridges are a vital element of the road network for the rugged, well watered terrain necessitates a bridge on average every five kilometres.

Therefore, permanent improvements were to be effected on one quarter of existing temporary bridge structures (such structures comprise almost half the total number of bridges). In paying special attention to the road network the government has been influenced by the presence of nearly half a million vehicles on the nation's roads, a number that is growing at an annual rate of 10-20%.

In 1970 the Philippines possessed only one passenger car per 130 persons; such vehicles overwhelmingly were concentrated in metropolitan Manila, which contained one third of all vehicles registered (see Fig. 38). Luzon, the largest and most developed island, and the one with the greatest extent of suitable terrain, accounts for some three-quarters of the nation's motor vehicles, and an even greater proportion of the non-commercial vehicles. In contrast, in certain of the pioneer areas of Mindanao, non-commercial vehicles are a rarity and usually consist of jeeps, the only type able to traverse the extremely rough roads.

Passenger traffic, therefore, is in the hands of the public bus companies. Some long distance carriers operate in Luzon and Mindanao but most operators consist of small companies each with half a dozen buses or flamboyant *jeepneys* (Manila alone possesses some 23 000 of the latter). Business is extremely competitive which means very low returns for operators. The result is an absence of fixed routes and schedules, dangerously high speeds, minimal maintenance, and overloading not only of passengers but also of cargo which frequently is carried.

Many handicaps stand in the way of road transport development in the Philippines. In most parts of the country, ease of access to the sea favours marine transport. Furthermore, the nature of the terrain focuses economic activity on coastal regions with the result that the road systems in the often rugged interiors are rudimentary. Even in areas generating substantial traffic, low standards of road construction and maintenance, coupled with the all too frequent violence of the elements, make for road systems noted for their inefficiency. Add to this a small demand for privately owned vehicles, especially outside the big cities, resulting from low personal incomes and high government taxes, and one is faced with a pattern of road transport totally inadequate for the country's needs.

Commerce

The vital importance to the economy of commerce is illustrated by the fact that it closely rivals manufacturing as the second largest employment sector after agriculture, fishing and forestry. In 1970 commerce accounted for no less than 16·0% of NDP (see Table 24). Also it has been growing appreciably; for example, from 1965 to 1969 the value of international trade increased by one third (exports only by 14% but imports by 54%) to give a grand total of £878 million (1970 data reveals a new pattern resulting from the devaluation of the *peso*). Parallel growth is believed to have occurred in the domestic sectors of commercial activity—wholesale and retail trade.

The expansion in international trade has brought with it considerable changes in its pattern. Most noticeable has been a reduction in the dominance of the United States as a trading partner. But that country still in 1970 was responsible for 44% of exports (c.f. 72% in 1949-50), and also 29% of imports. Much of the trade lost by the United States has been taken by Japan; in 1970, 40% of Philippines exports were destined for Japan (c.f. 6% in 1949-50), which in turn supplied 32% of imports, that is 10% more than the United States. The composition of trade, on the other hand, has altered little in recent years—the export sector is dominated by raw or semi-processed agricultural and mineral products (see Table 7), and imports consist mainly of machinery and transport equipment, base metals and foodstuffs (Table 30). Consumer goods, together with cereals after 1967, play a less significant, though still substantial, role as a consequence of efforts in the field of import substitution.

Import substitution is but one means by which the government is attempting to rationalise trade to harmonise with the pattern of economic development. Export expansion is the cornerstone of the government's policy, but so far it has had limited success, growth in the past being due to natural market forces rather than as a result of promotional effort. Similarly, attempts to find new markets , with the exception of Japan, have been largely unsuccessful. In an effort to avoid a return to the stringent licensing and exchange controls abolished in 1962, the government adopted in 1967 strong measures, notably tax incentives, simplification of export procedures and the creation

TABLE 30

Principal Imports, 1965-70 (£ million)

	1965	1966	1967	1968	1969	1970 (Est.)
1. Machinery other than electric	59.3	63.5	95.4	99.4	106.8	93.9
2. Transport Equipment	36.5	46.2	54.6	60.6	48.5	73.4
3. Basic Metals	32.8	35.3	44.2	45.6	48.9	60.2
4. Mineral Fuels and Lubricants	31.8	35.3	39.4	44.1	43.8	49.1
5. Electric Machinery, Apparatus and Appliances	20.2	15.1	19.7	25.3	28.4	24.7
6. Explosives and Miscellaneous Chemical Materials	10.2	12.9	15.2	16.0	16.9	20.7
7. Textile Fibres not manufactured into yarns	9.1	14.7	12.2	20.6	17.4 (est.)	16.8
8. Dairy Products	10.9	12.0	12.3	14.5	15.5 (est.)	14.9
9. Cereals and Cereal Preparations	39.8	22.2	34.4	17.0	18.5	13.1
10. Textile Yarns, Fabrics and Made Up Articles	7.1	12.9	13.4	18.2	16.1	n.a.[a]
Total	257.5	270.1	340.8	361.3	360.8	356.4
Others	81.6	88.2	102.1	191.9	131.6	97.0
Grand Total	339.1	358.3	442.9	553.2	492.4	453.4
% of Principal Imports of Total Imports	75.9	75.4	76.9	65.3	73.3	78.6

Source: Central Bank of the Philippines.

[a] Displaced from tenth position by chemical products and compounds (£15.3 million).

of an export co-ordination centre, in an effort to promote exports. In the import field, emphasis was placed on screening out non-essential products, i.e., those which can be produced locally or which are regarded as luxuries or semi-luxuries. As a result, a range of selective exemptions from the high level of import duties and taxes was established. In practice this system was thwarted by unrestrained consumer orientated expansion of the economy which culminated in the *de facto* devaluation of the *peso* in February 1970. This, coupled with the tough credit and budget policies adopted by the incoming administration (ironically re-elected in large measure by delaying such action until after the elections), by 1971 had succeeded in almost restoring the balance between imports and exports, though at the cost of drastically slowing down economic development.

U.S.-Philippines Trade. Success in rationalising international trade hinges on adjustment to the consequences of the 1954 U.S.-Philippines Revised Trade Agreement, otherwise known as the Laurel-Langley Agreement. This is designed to eliminate gradually the preferential trading rights established between the two countries in the days of the American Administration. Under the Agreement, Philippine exports to the United States were to enjoy a 40% margin of tariff preference until 1971, and a 20% margin from 1971 to 1974 at which date full U.S. duty rates would apply. Correspondingly, U.S. exports to the Philippines were to enjoy a 10% preference until 1974.

Products which together account for about one half of the total value of Philippine exports are affected by the Laurel-Langley Agreement. Of these, sugar is outstanding, representing as it does some four-tenths of the value of Philippine exports to the United States.

Exports of sugar to the Pacific Coast of America are on record from about 1800 but did not reach sizeable proportions until the second half of the nineteenth century. Then, sugar soon displaced abacá as the principal export commodity of the Philippines; during the last forty years of the Spanish regime sugar constituted 22-58% of the total value of all exports.[7] The United States was an important market in this period but it was far from achieving the dominance it was to attain following the establishment of free trade between the two countries by means of the Payne-Aldrich Bill Tariff Law in 1909. Thus, whereas in the decade 1900-1909 the U.S.A. took 23% of Philippine sugar exports, its share had risen to 56% in the following decade, and by the 1930s its share reached 99% of a trade volume some ten times the 1910-1919 annual average of 103 783 metric tons.

The flood of Philippine sugar into the United States had reached such proportions by the 1930s that a quota system was established to regulate supplies. The effect of the quota system on Philippine producers was to take away the incentive to increase production, for the domestic market was small and cushioned from free trade; Philippine sugar was not sufficiently competitive on the world market which was facing a situation of over-supply. This was despite the fact that Philippine production facilities were comparatively modern, the old *muscavado* mills having been replaced by *centrals* to cater for American demand for refined sugar.

The Second World War caused only a temporary disruption and soon the sugar producers were again in a position of strength, cushioned from external influences and rigidly controlled internally. It is only now that the complacency of the industry has been challenged, but many producers still refuse to accept that by 1974 they will have lost their trade advantages with the United States and will have to sell on an extremely competitive world market (for the domestic market, though growing, can absorb only part of total output). Philippine sugar producers are ill equipped to compete on world markets; in the past the sugar quota has been set at such a generous level that not infrequently the full amount has not been taken up—existing producers have been able to dispose of their output without trouble and domestic controls have thwarted the entry of new producers.

Not all sugar producers are blind to the need to adjust to the new trading pattern that is being forced upon them, nor are all so complacent that they have paid scant attention to new production techniques and let yields per hectare decline. Side by side with agitation to delay implementation of the Laurel-Langley Agreement (which could have short term success as powerful political figures are involved) a beginning has been made to restructure the industry. As will be demonstrated in the following chapter this could have important repercussions for the economic geography of the Philippines.

Aside from barter trade, which still persists in the Sulu archipelago, and the Manila Galleons, international trade has been a feature of the Philippine economy for barely a century and reached sizeable proportions only in the last half century. This growth in international trade revolutionised domestic commerce. Increasingly large quantities of imported manufactured goods were handled, reflecting the growth in purchasing power of Filipino producers of agricultural and other raw materials which found ready sale on world markets. The Chinese supplied the finance and entrepreneurial skill necessary to sustain the rising level of trade and inter-island marine transport facilities proliferated. In more recent years developments in the composition and orientation of international trade (the expansion of mineral exports by American companies, preferential tariff agreements with the United States, processing of raw materials prior to export, etc.) have modified the structure of wholesale and retail trade. Currently, the effects have been, or will be felt, of Japanese penetration of the market for manufactured and semi-manufactured goods; the ending of the Laurel-Langley Agreement; further sophistication of demand, especially by urban Filipinos; the government's import substitution policy; industrialisation; and improvements to transport facilities. Nevertheless, certain basic geographical factors remain dominant—the insular nature of the country, the low level of resource development, the strength of the Chinese middlemen and the Westernised outlook of many Filipino consumers.

In a nation such as the Philippines where manufacturing does not dominate the livelihood of the people, a preoccupation with wholesale or retail trade is characteristic of a significant proportion of those persons not engaged in primary production. Retail trade employment is approximately six times greater than wholesale trade employment. Of the latter, the largest single sector is devoted to commodity marketing (agricultural raw materials dealing) which has 24% of employees and over 40% of establishments (Table 31). General stores, with 65% of establishments, dominate retail trade, and *sari sari* stores are dominant within the sector (85%). Most other retail outlets are food stores in which the centrally located market predominates.

Perhaps the most important single commodity in domestic commerce is rice. This moves from the areas of surplus (the

TABLE 31
Selected Aspects of Structure of Wholesale and Retail Trade, 1961

	Nos. of Establishments	Nos. of Persons Engaged	Gross Receipts (£ million)
Wholesale Trade	6025	59 680	384
Agricultural Raw Materials Dealing	2449	13 959	77
Grains	358	2135	11
Copra and Coconuts	1090	5247	39
Abacá	128	1640	6
Tobacco	432	3171	15
Timber Products	227	1049	2
Retail Trade	121 860	281 703	310
General Stores	79 467	170 971	118
Sari sari stores	67 739	131 059	33
Food Stores	18 091	36 906	43
Meat and Fish markets	4798	8681	8
Fruit and Veg. markets	4530	7154	2
Grain stores	1555	3743	6
Tuba and native wine stores	3063	4789	neg.[a]

Source: Bureau of the Census and Statistics (Economic Census, 1961).

[a] neg. = less than one.

Central Plain of Luzon and the Cotabato Valley) to rice deficit areas, notably Manila, the Ilocos Region, and the Visayas. The Visayas, and particularly the Eastern Visayas, also need to import maize which is supplied by the Cotabato Valley and, to a lesser extent, the Cagayan Valley (the latter's surplus is destined usually for Manila). The recent Moro-Christian unrest in Mindanao was uncomfortably close to the areas of surplus in the Cotabato Valley and fears of an extension of the conflict were reflected in the price rises for cereals, always a highly sensitive indicator and one with great influence on political policy. Second to rice is timber, reflecting the elimination of adequate reserves close to centres of population. Today, Manila, Cebu and other cities have to rely on more distant forestlands, notably in Negros, Mindoro and Mindanao. Other construction materials figure prominently in domestic commerce but often, as is the case with *nipa* palm and bamboo, move only short distances, in part because of their unfavourable bulk/value ratio and also because supplies are more dispersed than good quality timber. Increasingly, timber has figured amongst the export commodities involved in domestic commerce, and recently it emerged as the principal item displacing coconut products. Abacá, sugar and also tobacco are handled in large quantities but minerals are usually exported direct. For the export commodities, movement is usually to Manila or the nearest port open to foreign shipping, at which point sorting, grading and processing may take place.

The domestic trade pattern for manufactured goods is less complex. The bulk of such goods originate in Manila either as a result of local manufacturing or as imports. Movement is thence to the inter-regional centres and then down through the hierarchy of trade centres (see Fig. 2). Leading products are food and beverages, clothing, fabricated metal articles, paint, soap and other chemicals, and paper products. Domestic commerce is of particular interest to the economic geographer as attention is focused upon respectively the basic objective of primary production—the creation of marketable commodities—and of the end point of economic activity—the satisfaction of consumer demand. A study of this sector makes it possible to gain considerable insight into the pattern and quality of life and livelihood of the average Filipino. It thus acts also as a prelude to the discussion of social development contained in subsequent sections. In order to achieve this objective a case study approach is essential involving in this instance presentation of the results of on the spot investigations of commodity marketing and retail trade in Hilongos—a small agricultural service centre in the Visayas. In microcosm there are presented two of the basic ingredients of life and livelihood common to many of the numerous agricultural communities which form the basis of Philippine economic and social geography.

Commodity Marketing. The municipality of Hilongos occupies the greater portion of the catchment of the Salog River whose floodplain provides the largest area of alluvial soil in South-Western Leyte. Four distinct production zones can be distinguished (Fig. 40): the beach ridges (principal occupations being fishing and growing of coconuts), the floodplain (rice, maize and coconuts), the foothills (coconuts, abacá, bananas and tobacco) and the interior (fruits, sweet potatoes, and cassava). Other products do not figure prominently in commodity marketing as they are the object of on-farm consumption.

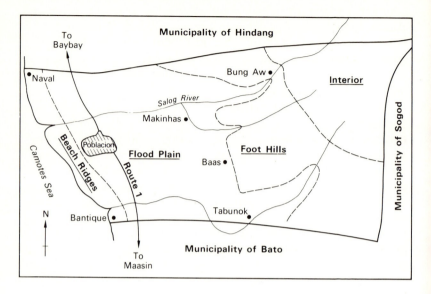

Fig. 40. Sketch Map of Hilongos Municipality, Leyte

The central location of *barrio* Makinhas within the municipality of Hilongos provides it with a larger hinterland than its rivals, Bung-aw, Baas and Tabunok. Each market is situated at the junction of the foothills and floodplain, thus providing interchange facilities for two different agricultural areas and also straddling a routeway to the interior which further extends the sphere of influence. Routeways usually are just tracks which rarely are adapted for any distances to motor vehicles; goods are usually transported by *carabao*-drawn sleds, horses and, as a last resort, the head and shoulders of the producer.

Marketing methods can best be illustrated by an examination of the movement of two dissimilar products from producer to consumer: fish, which is produced on the western fringe of the municipality yet is consumed throughout it; and abacá, which is produced primarily in the westernmost section yet is consumed almost entirely outside the municipality. To provide further contrast, fish is a highly perishable product but the sale of abacá can be delayed almost indefinitely.

Weather permitting, variable quantities of fish are landed each day at the numerous fishing villages fringing the Camotes Sea. Producers have to market their goods at once as no refrigeration or freezing facilities are available. On the beach, sale to fish merchants who await the fishermen's return provides the main marketing opportunity. For fishermen living near the *poblacion* of Hilongos its daily market offers an alternative opportunity. The fish merchant can sell directly on the daily Hilongos market or, if open, on the rural markets. On the Hilongos market competition from direct sales by fishermen is not great; most fishermen do not have the leisure or the initiative to dispense with the middlemen. In many cases, too, the fisherman has no choice as regards the sale of his catch, for the fish merchant often has a financial interest in it on account of his ownership of the fisherman's boat or, more commonly, the traps which provide the main source of fish.

Particularly when catches are good, recourse is had to sun drying or salting the fish for later sale to offset shortages of supplies of fresh fish which occur when stormy weather prevents the fishermen from putting to sea. There is a regular demand however for processed fish also and this especially is marked in those parts of the municipality distant from sea where supplies of fresh fish are irregular and expensive. The fish merchant may sell processed fish direct to customers at the Makinhas market but he also sells to the Chinese traders who take supplies of processed fish with them on their periodic visits to the interior and to other rural markets.

In contrast to fish there is not necessarily any urgency over the marketing of abacá. Nevertheless, abacá usually is marketed as soon as possible to offset the peasant's perpetual lack of funds. As with most agricultural commodities marketed in Hilongos municipality individual amounts marketed are small in volume for production areas are restricted and rarely mature all at the same time. As with fish, producers may have no freedom of choice as regards the trader patronised, for production usually has been pledged in advance in return for retail credit. In other cases, too, the output of tenants and sharecroppers usually is assigned by the land owner to a particular trader.

Abacá coming onto the market is purchased by the Chinese

traders who may by-pass rural markets by employing collectors to traverse the area on their behalf because individual output may be insufficient to warrant the producer making a special trip to market. Purchases of abacá are transferred to storehouses in Hilongos where they are graded and prepared for shipment. However, merchants may prefer to hold stocks until they regard prices as being favourable.

The above review of fish and abacá marketing reveals that there is basically no difference in approach. In each case the producer sells his output (less any retained for his own consumption) to a middleman who is responsible for the transference of the product to consumers in other parts of the municipality or beyond it. In the case of fish the middleman is bypassed to some degree by fishermen selling directly to consumers. Similarly, rural producers make small direct sales in their own villages or at local markets but the volume of such trade is not large. For both the fisherman and the farmer marketing flexibility may be restricted either by indebtedness or a lack of control over the ownership of means of production.

Retail Trade. For settlements such as Hilongos their *raison d'être* is trade and its retail form is of particular importance. It revolves around four separate institutions: the market, the general store, the specialist store, and the neighbourhood (*sari sari*) store. These distinctive features of the urban scene, individually and in conjunction, comprise important basic elements of the pattern of both economic and social geography in the Philippines. Consequently, treatment is more lengthy than that of preceding case studies.

Of paramount significance is the market, which is much more than a focal point of retail trade. Kirkup[8] suggests that the market and its environs portray:

'much the sort of scene one would have encountered in medieval times in Europe outside village churches or on the parvis of great cathedrals on important saints days and feast days. Close to the church, instead of a pageant, a grand outdoor concert was in full swing, and the audience at the concert and the congregation in the church seemed to be interchangeable— they kept going and coming from one to the other, to see what was new, what was going on. Thousands were listening to patter comedians and youthful pop singers with electric guitars: there was an extraordinary mood of carefree elation among these common people who though they might be poor, were enjoying the free treats of both church and concert'.

Under normal circumstances too, the market is a hive of activity, because uncertain, low incomes and the absence of refrigeration force many people to pay daily visits to their local market to replenish essential supplies. Not only does this maintain the typical all day, and all evening, congestion but it also makes essential the provision of comprehensive transport facilities. In consequence, the local market and the bus terminal are to be found in close proximity. In addition, the need for sources of supply to be easily accessible to the market means that it will have evolved on a main highway location and, where available, within reach of water and rail transport also.

Most markets are overlooked by the Roman Catholic church with whose interior the often devout Filipino is as familiar as with the interior of the market itself. The dimly lit, peaceful and wholesome atmosphere of the church contrasts vividly with the noise, the smell and the often badly maintained facilities of the market. Also adjacent to the market often are to be found one or more cinemas, for the Filipino is an avid picturegoer, though usually caring little for other arts, except dancing. Other forms of entertainment to be found in the vicinity of the market include the sidewalk musicians, fortune tellers and similar impromptu performers, the billiard hall, and less innocent but nonetheless well patronised activities, notably gambling, chiefly in the form of cockfighting and *mahjong*.

The daily visit to the market place thus can be as much a social as a shopping expedition. To many, and especially the numerous unemployed, entertainment is a luxury they can ill afford but in the cool of the evening the *plaza* provides an opportunity for a stroll or a chat. This relaxing and enjoyable pursuit will often be supplemented by the Filipino's favourite pastime, dancing, notably at *fiesta* time. On other occasions the promenade in the *plaza* may be enhanced, or disturbed as one's personal preferences dictate, by the loudspeaker-amplified exhortations of the local politicians.

The Hilongos market (Fig. 41), although not as grandiose as some, is typical in most respects. It comprises several permanent

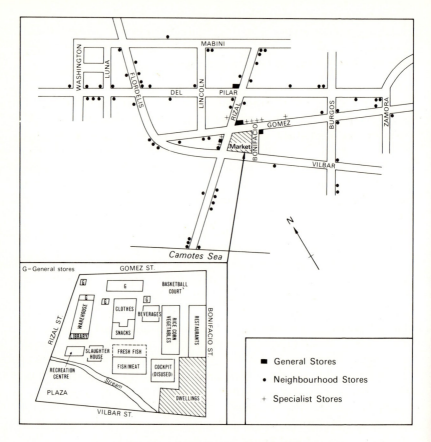

Fig. 41. Retail Trade Outlets in Hilongos *Poblacion*, 1967

structures, the most substantial of which is the warehouse in which many traders store unsold stock at the end of the day's trading. Such storage facilities are necessary as many market stalls are only partially enclosed, a concrete floor and a roof to provide protection from sun and rain being the basic amenities provided. Individual stalls, booths and shops rarely occupy more than forty square metres of low rental floor space. Stocks are minimal due to lack of capital and, in some cases, the day's wares may have been bought with the proceeds of the previous day's sales.

The north-east corner of the Hilongos market is dominated by traders who, by handling a variety of dry goods, are in direct competition with the adjacent general stores. Sales of other necessities of life dominate the remainder of the market area. One whole structure is devoted to rice, maize, vegetables and similar foodstuffs, while sales of fish and meat (the latter from the adjacent slaughterhouse) dominate the southern half of the market. Traders specialising in the sale of beverages are congregated in the south-east section of the market. A further substantial portion of the market is given over to restaurants and snack bars. The role of the market as a place of entertainment and recreation for the local populace is emphasised by the uses to which the peripheral areas are put. These include a small library, a recreation centre (billiards, *mahjong*, etc.) and the *plaza* which forms the focal point for festivals and political rallies. A further source of entertainment—the cockpit—is now disused as cockfighting is forbidden by law within 100 metres of a market place.

Hilongos in 1967 possessed five general stores all operated by Chinese. Within the market area a Filipino operated dry goods store has grown in status so that it can now be regarded as on a par with the general store. With the exception of highly perishable products the general stores stock virtually all items available in the market as well as monopolising the sale of luxury and semi-luxury items, and of hardware required by the community's entrepreneurs—sewing machines for the tailors and motor bicycles for the motor cabs, for example. For the bulk of the local population the range of goods provided by the general store encompasses virtually the entire spectrum of their demands[9] as the following indicates.

Only a score or two different varieties of tinned foods are stocked. Powdered or condensed milk and tinned margarine are prominent because no fresh milk is available and refrigeration facilities are lacking. Supplies of cereals, vegetables and fresh fruit are of poor quality as they are usually of local origin, often having been produced from unselected stock growing on haphazardly maintained land. The principal source of protein is fish; with rice or other cereals it dominates the diet to the exclusion of fruit and vegetables, an unexpected deficiency considering the scope for production offered by climatic conditions (such a state of affairs is widespread in the Philippines and now attempts are being made to redress the situation through the 'Green Revolution'—a self-help programme to encourage houselot cultivation). In a tropical climate beverages are consumed in large quantities. Soft drinks, coffee and cocoa are the most popular with beer and a local beverage (*tuba*), being the principal alcoholic stimulants. Sweetmeats and confectionery, too, find a ready sale. Local brands of cigarettes are the most popular, but solely on a price basis as their quality is much inferior.

Household requirements vie with food and drink as the focus of consumer demand. Low income levels and the tropical climate combine to give a simplicity of demand for clothing: dresses, shirts, trousers, shoes and hats being the chief items. Wood is in universal demand as fuel for cooking and, despite the recent introduction of an electricity supply system, lighting still relies heavily on such items as candles and kerosene lamps. Cooking materials reflect the local diet: garlic, onion, vinegar, soya sauce and the like are employed to flavour the various cereal, vegetable and fish dishes. Cheapness and an ability to stand hard wear are determining factors in the choice of cooking utensils, crockery and cutlery. Laundry aids are restricted in scope; soap is coarse and the limited demand for soap flakes confines supplies to small packs only.

The demand for personal requirements other than household remedies is irregular. Being non-essential items, demand is greater when the inhabitants are most prosperous, e.g. at harvesting time, or when local events are celebrated, e.g., at *fiesta* time. In general, items of personal hygiene are sought on a regular basis mostly by the well-to-do section of the community; in contrast, the younger elements in the community seek beauty aids. With literacy at a high level and education of the youthful population a major local 'industry' paper products sell well also.

With rising living standards the number of specialist stores is growing, having more than doubled in number in the period 1963-71. Most are located close to the general stores and the market, the focal point of consumer activity. Here, for example, are to be found establishments selling bakery products, pottery ware, and reed mats, simple basic products which usually are manufactured on the spot. Increasingly, more sophisticated products are being retailed and the range of establishments broadened to include petrol service stations, pharmacies, furniture stores, a farm supplies store, and a bank. In the 'suburbs' are to be found larger commercial establishments such as an ice plant, welding and motor repair shops, and builders' merchants. In 1971 Hilongos, as befits its role as an agricultural service centre, also held an agricultural show immediately to the southwest of the market, on a site reclaimed from swampland specifically for such functions and others designed to stimulate local agricultural expertise.

The distribution of neighbourhood (*sari sari*) stores reveals their significance in terms of numbers, though not in volume of sales (Fig. 41). Products most commonly stocked are basic items—eggs, bananas, beverages, cigarettes, candles, and laundry aids. Staple foods and perishable items normally are not stocked for the *sari sari* store cannot compete with the market in this sphere. The principal function of *sari sari* stores is to provide nearby residents with easy access to basic essentials, hence they are located usually at road intersections or centre sections of the longer streets. Thus, their nature will vary according to location. For example, tinned goods are more likely to be stocked in the wealthier residential areas and *tuba* will be more popular in poorer residential areas. Del Pilar street attracts a disproportionate number of stores as it forms part of the west coast highway; neighbourhood stores located there gain additional custom from passing travellers. Like most retail establishments neighbourhood stores are open from dawn to well past dusk, but, in this case, retail trade is essentially a part time occupation, the long trading hours being maintained by the continuous presence of one or more members of what is usually a large family. Goodwill plays a major role in their success or failure, particularly in their subsidiary role as a place to rest, drink and gossip.

B. Social Geography

Any developments in this sphere must take cognizance of the country's population growth rate which is amongst the highest in the world. Currently, it stands at 3·2-3·4% per annum, double that experienced in the Spanish era. More important, continued improvements in diet, public health facilities, the absence of epidemic diseases, continued resistance on the part of Catholics to birth control, and low rates of emigration suggest that growth may exceed 3·5% in the 1970s. Unless a combination of improved education and urbanisation produces a more sophisticated outlook and causes a reduction in the currently popular early marriages and large families, a population total in the vicinity of 60 million (double that of 1963) is possible by 1980. Success for economic development in the Philippines depends much on the nature of its society both now and in the foreseeable future. Temperament, social mores, nationalism and regionalism; these and other features of Filipino society can exert pressures which not even massive amounts of economic aid can resist. Economic development may create a more sophisticated society but the future welfare of the nation depends primarily on basic attitudes towards life not livelihood. Such attitudes exert a considerable influence on the pace, the nature and the location of Filipino efforts to secure a greater control over the environment.

The self-help programme inaugurated in 1956 by the Presidential Act on Community Development (PACD) provides a useful indicator of social attitudes. The objectives of PACD are to:

(a) promote the development of self-government in the rural *barrios*.
(b) maximise the productivity and income in the rural areas through accelerated self-help development of agriculture and other industries.
(c) expedite the construction, largely on a self-help basis, of approach roads which will connect all *barrios* with principal highways.
(d) expand governmental services to *barrios* to an extent commensurate with those now available in the *poblaciones*.
(e) promote better coordination of government services in all levels.
(f) improve through self-help rural facilities for education, water supply, irrigation, health and sanitation, housing and recreation.
(g) improve the educational/vocational opportunities for the adult population of the rural areas.

(h) increase citizen awareness and action with respect to the enforcement of laws on farm tenancy, usury, labour and other subjects.
(i) take all other steps possible to improve the morale of the *barrio* residents and to strengthen their sense of participation in the economic and social life of the nation.

PACD is one quasi-government body which has suffered little from over-ambitious objectives, waning enthusiasm, malpractices, and operating inefficiencies. Field operations and programme assistance are channeled through local development councils. These councils are co-ordinating institutions whose membership includes representatives of technical departments operating in the area. Implementation of PACD operations at this level is in the hands of *barrio* development workers each of whom is responsible for up to three *barrios*. By 1970 the programme had contributed directly to 47 834 projects—irrigation schemes, waterworks, schools, health centres, etc.— valued at about £17 million. PACD stimulation and guidance contributed to a further 596 820 projects valued at nearly £32 million. Its total field force exceeds 2000 persons serving some 5000 *barrios*.

In outlining the economic geography of the Philippines above, certain aspects of a socio-economic nature have been discussed. Into this category fall employment, the distribution of wealth and, more particularly, social patterns of retail trade. Still requiring attention are various fundamental geographical features of Philippine society. Their current features have been outlined in a previous chapter (The People). At this stage, therefore, emphasis is placed upon the highlighting of the more volatile elements. Even for a nation in a state of flux such as the Philippines, social change is slow and its impact on the country's geography even slower. Nevertheless, developments in certain sectors not only can have significant short term implications but also can initiate more far reaching long term effects. In the following pages an attempt is made to examine selected developments whose geographical consequences could be profound. From a sociological point of view others might deserve pride of place but developments leading to increased control by man over his own nature are secondary in importance from a geographical point of view unless they involve place as well as people.

The Rural Scene

In considering the current social geography of the Philippines, rural society needs to take pride of place in view of agriculture's intimate involvement in the present way of life of the majority of the population. In this context, land reform is paramount; indeed, apart from agricultural resettlement (transmigration), no other appears to be of consequence in the foreseeable future. Agricultural development schemes can be expected to modify the agricultural way of life by, for example, accelerating the drift from the land through mechanisation, reducing parochial attitudes through better communications, and improving living conditions through drinking water supply schemes. However, without land reform not only will agriculture be characterised by inefficiency, whatever the intensity of investment and extension services, but also the agricultural worker will be deprived of a just return for his labour and in consequence will not adequately participate in the benefits accruing from overall economic and social development.

Before discussing land reform it is advisable to examine more closely the implications of demographic features for rural development. The cities of the Philippines may suffer increasingly from overcrowding but it is overcrowding in the countryside, containing two-thirds of the nation's population, which is the root cause of many of current socio-economic problems. Survey data[10] indicate that nearly half the rural population consists of children below fifteen years of age. The children are the product of exceptionally high rates of fertility: average completed family size exceeds six in number (double that of rural areas in developed countries) and the proportion of childless women with completed fertility is less than 4%, compared with over 10% in industrialised countries.

Over 6 million of the 8.3 million rural workers are engaged in agriculture. There thus is revealed the often neglected fact that about one quarter of rural inhabitants are engaged in other activities—construction, manufacturing, commerce, transport, government and domestic service, etc. Most of such rural inhabitants generally are to be found in the smaller *poblaciones* (population less than 2500 persons) and *barrios* (less than 1000 persons). Of greater significance is the fact that nearly one third (1.9 million) of agricultural workers are employed on average for less than forty hours per week. When voluntary part time workers are excluded there remains a hard core of 0·9 million persons, or 14·9% of all employed agricultural workers, who expressed a desire for additional work. Furthermore, another 0·9 million persons in the fully employed category also wanted additional work. On this basis therefore no less than 29·8% of the agricultural work force is under-employed.

The chronic lack of job opportunities in agriculture is largely responsible for the lack of skill of rural labour, its reliance on meagre savings, lack of response to innovation, and generally traditional and parochial outlook. In turn, the lack of job opportunities in no small measure stems from ever increasing population pressures which urbanisation and rural resettlement only partially can offset. As indicated above, exceptionally high rates of fertility exist in rural areas and when these are associated with a general lack of birth control policies it is inevitable that rural population growth will absorb all those resources which might otherwise be used to increase per caput physical and personal capital. Dissipation of such resources, not only through population growth but also through undesirable systems of land tenure, is the root cause of the depressed state of Philippine rural society today.

Land Reform. The desire for land ownership and a release from economic bondage have been primary causes of social unrest in the Philippines since the time of the Spanish. Despite the well-intentioned efforts of the American Administration, and subsequently of President Macapagal, little positive action has been taken. Unrest has not reached a scale which would cause pressure groups representing the land-owning élite to reconsider their determination to maintain the near-feudal structure of Philippine agricultural society. Indeed, if anything, the successful implementation of land reform measures, miniscule as it is, is more than offset by increases in farm tenancy in other parts of the country. Certainly, the position is now far worse than at the turn of the century. Then, almost immediately following the overthrow of the Spanish, the tenancy rate for the whole country was 18% (excluding farm labourers) but in the 1960s the comparable figure was over 50%.[11] Taking farm labourers into account some two-thirds of the rural work force were landless, not to mention the numerous unemployed.

The critical area in terms of land tenancy is the Central Plain of Luzon (see Fig. 32). Here, on the largest and most fertile area of agricultural land in the country, tenancy rates are well above the national average, reaching 90% in Pampanga. At the other end of the scale, about one half of the agricultural land is owned by 0·5% of the population, with the Catholic Church figuring prominently as a landlord. The combination of land alienation, subsistence agriculture and intense population pressure existing in the Central Plain of Luzon has offered an ideal breeding ground for socialistic and communistic movements. Those of recent origin gained considerable momentum after the Second World War, as a result of the stimulus provided by the granting of independence and experience of guerrilla war against the Japanese. Using the swamps of Pampanga as an easily defended base, the Hukbalahap or the 'People's Liberation Army', popularly known as the Huks, gained the tacit and sometimes willing assistance of the local population to become a force with which to be reckoned.

Before and after ascending to the Presidency, Magsaysay contained and then reduced the power of the Huks by aligning government policy in favour of the lower classes to a previously undreamt extent. Magsaysay's policies did not, however, survive his death with the result that Huk agitation was revived, but not revitalised, thanks to at least lip service being paid to land reform coupled with improved efficiency on the part of the law enforcement agencies. The resultant stalemate has caused the Huk movement to lose much of its revolutionary enthusiasm and most observers consider that it has to become equally concerned with a wide range of criminal activities, paralleling and probably connected with those characteristic of nearby Manila, as with politics. The 1970s, however, have seen the emergence in various parts of Luzon of the New Peoples Army (NPA) which appears likely to take over the politically militant role relinquished by the Huks. Irrespective of current motives, it remains a fact that the ability of the Huks and the NPA to flout the law when it suits them is only made possible by the assistance of many of the local population for whom they represent the sole means by which they can hope to achieve agrarian reform.

The local population thus is evidently not impressed by efforts at land reform and development in the Central Plain of

Luzon to date. On paper, government effort in the area has been substantial, culminating in the Agricultural Land Reform Code. Foreign aid organisations, notably the Philippine Rural Reconstruction Movement, also have been active. Legislation has been spearheaded by land redistribution programmes sponsored by Presidents Magsaysay and Macapagal which have been supplemented by technical and civic programmes designed to enhance the tenant's role in the production process and to improve in other ways rural community life. More recently, the entire Second District of Pampanga was designated a special land reform district under the Land Reform Authority and the Central Luzon Development Programme has been established to accelerate social and economic reform in the region as a whole. American aid, both in the form of money and personnel, has been provided to make available *inter alia* cheap credit for farmers and to supervise the growing of new, high yielding, strains of rice.

In terms of visible progress results have been disappointing. Much legislation suffers from inadequate preparation with the result that the landowners, with their better education, are able to exploit the various loopholes or else take advantage of the red tape provided by the inefficiencies of public service procedures. Finance available, too, has proved totally inadequate, including that supplied from foreign sources. The Agricultural Credit Administration (ACA) set up by the government to finance small farmers so far has received only a fraction of the funds (over £40 million) called for in the Land Reform Code. Equally small amounts of money are available so far for loan by the American government's Agricultural Guarantee Loan Fund. In contrast private enterprise banks have made great strides in the area but some are owned by landowners to finance their own operations and the others generally require land titles as collateral. Even the lessee who rents land, recently released from tenancy by the Land Reform Code, cannot borrow money. Finally, personnel problems have arisen in that the more capable of the local administrators also are engaged actively in combatting Huk and NPA activity. As a result, they frequently become victims of terrorist atrocities. It should not be forgotten also that conditions in some areas are such that local administration is in practice in Huk hands. Consequently, any success in rural development is credited by the local inhabitants to them and not to the central government.

The socio-economic conditions under which rural inhabitants exist are such that it is little wonder that a 'revolutionary' rather than an 'evolutionary' solution is embraced. In detail great local variation exists amongst those who would benefit from agrarian reform because land tenure and associated arrangements have acquired many permutations and combinations over the years. In terms of end result, however, a near identical state of affairs exists as exemplified by the following case study of two farmers from Laguna Province.[1,2]

One farmer, Miguel, in 1960 rented 2·30 hectares of unirrigated land; the other, Juan, rented 2·78 hectares on 0·70 hectares of which two rice crops were possible through pump irrigation. Both Miguel and Juan cultivated vegetables on nearby hillsides, and raised pigs and chickens for on-farm consumption. Each landholding was insufficient to provide full time employment all the year round so additional income was secured from helping other farmers transplant and harvest their rice (tasks for

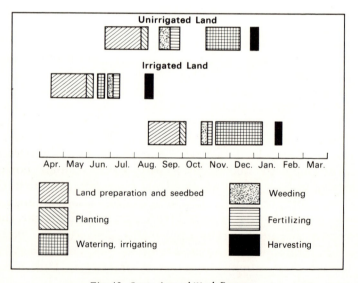

Fig. 42. Juan: Annual Work Pattern

which Miguel and Juan also required assistance at the appropriate time). With two separate crops of rice, Juan was able to devote twice as many working hours on his own holding as Miguel (Fig. 42). Additionally, Juan lavished more care on his crop, including the application of fertiliser.

Despite the significant differences in holding size and quality and farming attitudes, as regards personal benefit Miguel and

TABLE 32

Miguel and Juan: Disposal of Rice Crop and Off Farm Remuneration (kg of rice)

	Miguel	Juan
Rice Crop		
Outgoings		
Landlord's share	660	2046
Planter's share[a]	396	1232[b]
Harvester's and Thresher's share	308	972
Seed	88	176
Creditors	638	2046
Total	2090	6472
Total Yield	2112	6472
Net Return	22	Nil
Off Farm Remuneration		
Outgoings		
Creditors	704	453
Remuneration	880	673
Net Return	176	220
Total Net Return	198	220

Source: D. J. Dwyer, 'Irrigation and Land Problems in the Central Plain of Luzon', *Geography*, vol. 49, 1964, p. 242.

[a] Advanced by landlord and others to cover household expenses prior to harvest.

[b] Including fertiliser and petrol/oil for irrigation pump.

Juan were equally poverty stricken (Table 32). In Miguel's case almost all his rice crop was required to settle various debts and past expenditures while over three-quarters of his off-farm renumeration was similarly disposed of. In total, less than 200 kg of rice could be retained for on-farm consumption, a totally inadequate amount and one which ensured another year of indebtedness. Juan, though producing far more rice, was able to retain none at all once debts and past expenditures had been

settled. Debts to the village shop took some two-thirds of off-farm remuneration leaving his family with just 22 kg of rice more than Miguel as a reward for the year's labour.

Miguel and Juan, like over half the other farmers in the Central Plain of Luzon and adjacent areas of predominantly tenure farming, create little or no marketable surplus of rice.[13] Permanent indebtedness is thus the rule and the fear of incurring even greater debts inhibits technological innovation. For such farmers the green revolution caused by HYV rice often only will occur if initiative is taken by the landlord.[14] But, more often than not, the landlord is satisfied with the *status quo* for, despite the banning of share cropping, it persists in disguised forms thus adding fuel to agrarian unrest; in certain cases of sub-tenancy three-quarters of the gross harvest may be required to meet rental demands.[15] Unless the Agricultural Land Reform Code can eliminate quickly such abuses, revolution of a colour other than green could result.

Virtually the only practical large scale rural development of note has been the introduction of the so-called HYV or 'miracle rice' strains whose increased yields have almost made the country self-sufficient in rice. Lack of credit, however, has prevented its benefits reaching all rice farmers[16]; Miguel and Juan, for example, probably persist in their traditional way of life and livelihood. Paradoxically, it is possible that this development will hinder efforts at land reform. The higher productivity created by these new strains has made profitable large scale farming operations requiring large inputs of capital to finance mechanisation, irrigation, weed control, etc. Such inputs only can be supplied by the bigger landlords and can succeed only if a landlord and labourer relationship replaces that of landlord and tenant. Furthermore, with the spread of rice self-sufficiency in South-East Asia as miracle rice production expands, a fall in prices can be expected thus increasing the difficulties of the small landholder. With only the most efficient farming units of current minimum economic sizes[17] likely to be viable in the long run (a trend likely to apply to other agricultural products as well as rice) land reform policy could create a new grouping of impoverished rural inhabitants who might lose their new found freedom of land tenure as economic pressures forced them into a new debt cycle.

Short term prospects in land reform thus are likely to be

characterised by frustration and sporadic small scale progress; in the long term, even if not disrupted by economic forces, the rate of progress can be expected to be only marginally faster. One proviso needs to be made, in the light of the scope for increased awareness by tenants of their plight. In this context one can do no better than quote from the detailed study of land tenure in the sugar lands of Negros made by Schul:[18]

'A new generation of workers, however, will not judge its situation on past conditions which they have not experienced but rather will judge in terms of new aspirations. The seeds of discontent may find a receptive soil as the landless labourers become more conscious of the outside world and its values as developed by organised unions, published materials; and possibly, most important of all, the radio.'

Certainly, Huk-type organisations recently have been reported in Negros where even the landowners admit that one in five sugar workers are not paid the wage level to which they are entitled. But, in general, the apathy of the agricultural worker persists.

To conclude, one must emphasise a point all too often forgotten by the proponents of land reform. For agriculture to make its proper contribution to economic growth it needs to be motivated by the criterion of high productivity based on large scale mechanised operations. In this respect, land reform as practised in the Philippines would appear motivated more by social than economic considerations. In terms of agricultural economics alone fragmentation of *hacienderos*' holdings could be a retrograde step in the light of anticipated developments. Once created, viable small holdings are not necessarily permanent; even the pioneer lands of Mindanao exhibit this trend despite hopes that the settlers would not reproduce the conditions operative in their former home regions but instead would become land holders on economically viable units. In Southern Mindanao, according to Wernstedt and Spencer,[19] 'on the pioneer fringe there is a strong tendency for the new settler to slip back into the position of a tenant on a too small holding after only a short period of occupancy on his new land, even when open and available land is to be found locally.' Economic and technological factors no doubt contribute to this state of affairs but it is suggested that cultural factors play an important role also. The latter, too, loom large in overall land reform policy—the relative merits of 'land for the landless' or 'food for the cities' are likely to be argued about for many years to come.

Transmigration. As used in an Indonesian context, this term usually applies to the unsuccessful efforts made to move people from densely populated Java to relatively sparsely peopled areas of Sumatra or other of Indonesia's Outer Islands. In the Philippines the problem soon will be less one of persuading people to move than one of finding land on which they can settle. The focal point of pioneer settlement—Mindanao—is rapidly running short of new land suitable for agricultural settlement and by the late 1970s the pioneer frontier throughout the Republic may have been pushed back to its ultimate theoretical limit.

It is not easy for Filipinos to comprehend that Mindanao is becoming a crowded spot, agriculturally speaking. Mindanao in the years of the American Administration recorded an inmigration totalling approximately 0.7 million persons.[20] Three-quarters were absorbed either by the north coast provinces or by Davao. In the post war period (1945-60) a further 1.25 million migrants took up residence. More recent reliable statistics are not available but there appears to be little doubt that this increased tempo generally has persisted, for 1970 estimates of total population continue to assume that the islands' overall growth rate has remained at a level nearly 50% greater than that of the national average.

The massive influx of people remains unevenly spread. Areas absorbing most people are the frontier regions of the interior and the south; the relatively long settled areas along the north coast have proved least attractive to immigrants. Leading destination of immigrants is Bukidnon Province, nearly 80% of whose population in the period 1945-60 could be attributed to inmigration.

Focal points of pioneer settlement have been the government sponsored projects. The direct contribution of government resettlement schemes has not been high, reaching in Bukidnon a maximum of one fifth of all settlement. The value of the schemes lies in the resultant publicity which has stimulated independent movement. Much of the latter, too, involves relatives and friends of government settlers. Encouragement to independent migration also has been given by government road construction works. Pioneer agricultural areas opened up by this method include the Agusan Valley, the Allah Valley, the Kapatagan Basin[21] and the Digos–Padada Lowland. Ease of access has proved an effective remedy for fears of malaria and Moros, the two most common reasons cited by would-be immigrants for not leaving their homeland for Mindanao.

By comparing data for the 1948 and 1960 Census, a detailed analysis of migrants' homelands is possible.[22] Nearly 150 000 persons moved the few kilometres from Misamis Occidental to Zamboanga del Sur. This intra-island movement represents 40% of the internal relocation of Mindanao's population, which principally involved movement from established areas of settlement to frontier areas. Inter-island movement, however, contributed more than three times the numbers involved in intra-island movement. Of the 1·24 million persons involved, the Visayas contributed 86·5% (30% from Cebu alone) and the Central Plain of Luzon and Ilocos together a further 7·5%. Migrants from outside Mindanao exhibit a certain regionalism in their preference for settlement: Cebuanos favour Davao and Zamboanga del Sur, Leyte born migrants favour Surigao, while migrants from the Western Visayas and Luzon prefer Cotabato. Cebuanos' preferences are influenced by long established trading links, those of Leyte migrants by accessibility, and the attraction of Cotabato stems from the concentration of government assisted schemes there which have a greater appeal to Filipinos having limited prior knowledge of the island.

The Visayan's mobility has led to the speedy colonisation of Mindanao, much to the chagrin of the Moros whose mainland sphere of influence now largely is confined to Lanao del Sur. In parts colonisation has been too swift: ill considered forest clearance, the importation of traditional but undesirable tenure patterns, and arbitrary disregard for the socio-economic framework established by the Moros and pagan rural inhabitants have weakened the foundations which have been laid for the island's future rural development. Fortunately, however, the rural sector should not be the sole basis for the development of Mindanao if current trends towards industrialisation persist.

The Urban Scene

Moving from the rural to the urban scene, the pattern of Philippine social geography takes radically different forms. Here, the emergence of an industrialised society has brought with it many features of more developed nations such as slums, a lessening of kinship ties, and the commuter. Filipino social attitudes have minimised certain developments and accentuated others; thus, kinship ties remain relatively strong but low living standards, coupled with climatic conditions, make for exceptionally bad slum conditions. From a geographical standpoint, efforts to improve the urban environment follow similar lines to those in other parts of the world although the money and services available are, by virtue of the less mature nature of the community, not always used as wisely.

Despite similarities to other parts of the world it is desirable to examine more closely, from a geographical viewpoint, certain features of Philippine urban society. The residential pattern, and the suburban pattern in particular, is of special interest for not only does it reflect the unique characteristics of the Filipino's urban environment but also it represents a feature of the landscape likely to expand markedly in areal extent in the years to come. There is another feature which merits attention for it exemplifies a significant change that is taking place in the structure of Philippine society, and one which can be expected to have substantial long term implications. This is the introduction of Filipinos into top management, or 'Filipinisation' as it is more commonly known, which symbolises the impending emergence of a truly independent nation. In practical terms also, Filipinisation has important consequences for the pattern of economic development and for the strengthening of that backbone of democratic society—the middle class.

Suburbanisation. Settlement patterns in the Philippines were noted for their immobility in the Spanish era, littoral movement associated with the flexibility of the nucleated *barangay* unit having been stabilised by the colonial government for political and military reasons. Increasing population pressure now has caused a more flexible situation in which improved forms of communication and industrial development play important roles. Thus, for example, an improved road network has favoured movement to the interior instead of along the coast, and also made ribbon development commonplace. The dominant movement of population, however, and one far exceeding the much publicised opening up of the pioneer lands of Mindanao, is to the cities to take advantage of job opportunities created by industrial development. In this respect, the Metropolis of Manila is, of course, outstanding.

Suburban patterns in Manila have been slow to change since the Second World War despite the stimulus provided by war damage and the influx of people from rural areas. Nonetheless, significant changes now are taking place. Such changes are of special interest for eventually they can be expected to become apparent, in a modified form, in Cebu City and other of the country's intra-regional urban centres. In this context, note

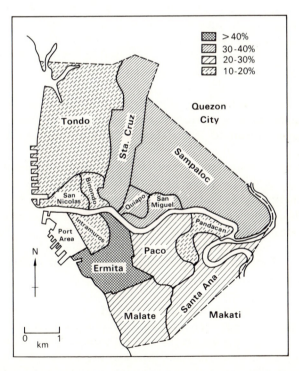

Fig. 43. Manila City. Percentage of Population in each district with higher education

should be taken of the probability that in the two decades 1958-77 the urban population of the Philippines will have risen by over 10 million people to a total of 18 million. By then, metropolitan Manila will have a population of perhaps 6 million of which some 40% will be a consequence of immigration.[23]

In Manila proper four distinct suburban quarters can be distinguished (Fig. 43). Within each, the basic urban scene exhibits a variety of modifications largely governed by the living standards of the population. However, such standards should be viewed in their true perspective. Throughout, overcrowding is a problem for the area contains one half of the population of the metropolis in less than one tenth of its built up area. Furthermore, whereas an annual house building rate of twelve per 1000 families is considered essential to cope with current overcrowding the actual building rate is nearer two per 1000 families.

The Chinese district of Binondo is the core of the 'inner city': an area to the north of the Pasig River characterised by a mixed population, apartment houses of various forms, and an above average provision of services such as flush toilets and piped water.[24] The suburbs surrounding this area are more typically Filipino. Three categories can be distinguished: slum areas focussed on the Tondo district, lower middle class residential areas typified by Sampaloc and Paco, and the upper middle class area to be found exclusively south of the river and attaining its ultimate in the Ermita district. In the slum areas the population is dominated by unskilled labourers and fishermen living in traditional houses built of *nipa* palm and bamboo. The lower middle class inhabitants dwell in single family houses in which timber is the chief constructional material; their access to services, their educational standards, their radio ownership, etc., represent the mean for the city. The better off inhabitants of Manila proper exhibit above average standards in these fields, dwell in substantially built homes, and are the most Westernised and sophisticated of the city's population.

The above polarisation of the four basic elements of the inner metropolitan society has been in existence for many years. It has resisted change because the attraction of their respective individual focal points—the Central Business District, the Port, the Pasig River factories and Malacañang Palace—has been unchallenged. Recently, however, the emergence of additional foci, improved transport facilities, and action by the local authorities have contributed to a significant shift in the pattern of suburban residence. One of the most potent of additional foci has been the growth of tertiary employment opportunities in Quezon City, associated both with establishment there of the University of the Philippines and progress in transferring government departments to what is the official capital of the Republic. Rivalling Quezon City in opportunities for tertiary employment is Makati, such demand being a function of that suburb's industrial growth which makes it and the adjacent suburb of Mandulyong the prime focus for new employment in present day metropolitan Manila.

Makati is atypical, yet at the same time it illustrates many facets of urban life in the Philippines and, more important, serves as a symbol to Filipinos of their future urban environment. The 2500 hectares of Makati muncipality represented, in the expansionary years following the Second World War, the only sizeable area of undeveloped land in close proximity (twelve kilometres) to downtown Manila. Even more significant, over half the area was owned by a single enterprise—the Ayala Corporation—which resisted the temptation to dispose piecemeal of its holding in response to the pressing demands for industrial, office and residential space resulting from the wartime destruction of the downtown area and the post-war influx of people to the capital. Gradual and controlled growth was the order of the day (as a result, the total value of property in the area rose 140 times in the period 1948 to 1965). Initial development involved the rehousing of squatters by the provision of low cost lots, the establishment of industry to provide jobs for the former squatters, and the provision of services, roads, schools, shopping centres, etc., to serve the industrial community. Special emphasis was placed on catering for business executives, both those employed by the individual firms established there and those employed by the commercial and financial organisations subsequently lured to Makati by another facet of its development in the form of modern, well planned multi-storey office blocks. The result was Forbes Park, unique at the time of its creation in that residents enjoyed paved streets, reliable water supply and sewerage systems, elaborate landscaping, stringent zoning of land use, and efficient street lighting, not to mention a golf course and polo ground.

Established metropolitan foci have lost relatively little ground mainly because improvements to transport facilities have been meagre and have been largely negated by the increase in traffic. With fierce competition keeping public transport charges exceptionally low and with the growth of vehicle ownership increasing the mobility of executive and managerial staff, the attractiveness of downtown Manila as a source of employment has been maintained despite the competition from Quezon City and Makati. Ever greater densities both in the daily work force and in residential accommodation have resulted. Thus expansion here has involved increasingly a vertical element. As yet, however, it is only on a modest scale, particularly as the area's susceptibility to earthquakes acts as an inhibiting factor.

Nevertheless, improvements in the field of transport have encouraged suburban development beyond the boundaries of Manila proper. However, the desire to get away from its crowded conditions has been only in part voluntary. Factors such as the provision of cheap housing by local authorities, a crackdown on excessive slum development such as in Intramuros, and the attraction of Quezon City, Makati and subsidiary employment foci have combined to create important centrifugal forces strong enough to break the ties of convenience and kinship which until now have bound many people to Manila proper.

Filipinisation. In most developing nations nationalism, in one form or another, represents an important element of social geography. The Philippines is no exception; established patterns of society are being modified, especially in the urban areas, as a result of both naturally occurring and also enforced replacement of expatriates, including the Chinese, by Filipino citizens. This development is but a stage in the emergence of a society capable of assimilating external influences instead of being overwhelmed by them, a society whose character is truly representative of the fundamental geographical features of the archipelago.

The underlying current of anti-Americanism prevalent in Philippine society primarily expresses itself in words rather than deeds, and any practical consequences usually fall far short of their exponent's original intentions. Yet, there is one development which quietly has been gathering strength and which has major long term implications for the future of Philippine society. This is the increasing share being taken by nationals in top management, thus displacing expatriates who, because of the nature of foreign investment are usually citizens of the United States (approximately three-quarters of foreign investment is American in origin and this represents about one fifth of total investment in the Philippines). One of the most important appointments in recent years has been the installation of Geronimo Velasco as President (now Chairman) of Dole Philippines. Not only is the company one of the more prominent of the American financed corporations but Mr. Velasco was appointed on a strictly professional basis. His ability had in fact been recognised earlier for in 1965, when with Republic Glass, he was cited by *Asia Magazine* as one of the top executives in the Far East. Increasingly, related appointments have been made in recent years as foreign corporations have come to appreciate the presence of a growing pool of indigenous management talent (these include appointments with more than just national significance: Theo H. Davies & Co. (Far East) Ltd.—a trading organisation with many Asian interests—has a Filipino both as President and as General Manager). In no small measure such talent has been created by the corporations themselves through their staff training schemes. The success of these schemes having been established, it can be expected that they will become of greater significance in the future thus ensuring that the trend towards Filipinisation of management will continue to accelerate.

Paralleling the replacement of expatriates has been a broadening of managerial skills in Filipino owned enterprises. In part this has been stimulated by the injection of executives benefitting from the training programmes of foreign companies. More important, however, have been the activities of a handful of self-made millionaires who have made the most of opportunities arising in growth sectors as, for example, the timber industry, real estate, and the Iligan steel complex. These persons represent a new element in Philippine upper class society—men with no aristocratic or landowning background but well endowed with determination, drive and business acumen. These *nouveaux riches* themselves are unlikely to make the impact on society they have had on the economy but the example they have set and the opportunities they have created for lesser men will have profound consequences.

•Finally, there has been a determined effort by the government to minimise the influence of the Chinese by Filipinisation of those sectors of the economy they dominate. Such measures have not always been successful in practice as the Chinese have successfully resorted to the use of 'dummies'. Nevertheless, the government's policy has further widened the scope for direct Filipino involvement in the economy which, although not always beneficial in financial terms, cannot but strengthen the country's social structure.

The impact of the government's Filipinisation policy can be illustrated by reference to those most vital of commodities—rice and maize. Nationalisation of the rice and maize industry occurred in 1961 when the Rice and Corn Board was created with the responsibility for the successful 'Filipinisation' of the industry. The first task of the Board was to weed out all the aliens who were already in the industry, numbering 6109, more than half the total participants. These aliens, mostly Chinese, were given only two years to liquidate their business. In order not to deplete the industry, some £10 million was made available. Half was allocated for assistance to the natural born Filipinos engaged in the industry to finance their operating capital and the acquisition and construction of facilities for processing, warehousing, milling, transportation and other facilities needed by the industry. The other half was provided to finance production, farm improvement and development of Filipinos engaged in the actual farming and production work. This money was not always well spent but the Filipinisation activities of the Board have meant that all the 10 000 or so millers and 500 or so wholesale organisations and warehouses engaged in the industry now are registered in the names of Filipinos.

The Political Scene

Where Philippine social geography is concerned politics cannot be disregarded. Most political activity either is not translated into practicable terms or else has negligible direct geographical significance. Nevertheless, every so often a political issue has geographical implications; for example, the creation of new provinces, anti-Chinese legislation, import substitution measures, etc. Many such issues have been assessed in the preceding pages but outstanding by its absence has been one whose implications are of principal interest to the political geographer —the Sabah issue.

The Sabah Issue. This subject has frequently been in the forefront of the Philippine political arena in the 1960s although at the end of the decade it appeared to have lost its sense of urgency. Nevertheless, it is likely to remain a volatile element which, although primarily of concern only to the Philippines and Malaysia, has implications which both involve broader issues in the field of foreign relations and also could lead to significant changes in the life and livelihood of those parts of the archipelago closest to the disputed area—South-Western Mindanao.

Sabah, formerly known as North Borneo, has had a varied past. It has been claimed by various states or collections of states. In the nineteenth century the two primary contenders for sovereignty were the then independent Sultans of Brunei and Sulu. The Sultan of Brunei achieved his recognition as a result of his original effective ownership and occupation of the territory which, by circumstance, he was forced to relax in practice if not in theory. In contrast, the Sultans of Sulu never effectively occupied the territory, although they encouraged the activities of traders and pirates in the area; and with the latter they occasionally enthusiastically joined in slaving and pillaging expeditions.

The Sultans of Sulu based their claim to Sabah on their intervention early in the eighteenth century in the affairs of their sister Sultanate of Brunei in one of the inevitable quarrels of succession. Their claim was that North Borneo was awarded to them in return for their military aid. However, no deed or treaty of cession now exists to support this claim and successive Sultans of Brunei have denied that a cession of North Borneo to Sulu ever took place.

Malaysia claims its right to ownership from the chain of events initiated with the Sultan of Brunei's grant of Sabah to the British North Borneo Company in 1877. The company ran the territory for a while until it was later declared a protectorate by the British Government. North Borneo became a Crown Colony after the Second World War and subsequently gained independence as part of newly formed Malaysia.

The Philippines rests its claim on the assertions of the Sultans of Sulu who were conquered and their land annexed by the Spanish Government in 1878, who were in turn replaced by the United States, and ultimately, by the independent Republic of the Philippines. The Filipino claim presupposes that the Sultans of Sulu were in fact the legal owners of North Borneo in succession to the original owner, the Sultan of Brunei, who granted it to them for services rendered. The Filipinos allege that the document purporting to grant the territory to the North Borneo Company was not in fact a grant or cession in the strictly West European sense of the word which transfers the property for all time to another. They allege that it was not in fact a grant in perpetuity but a lease; therefore, sovereignty over the territory could not have passed to the British North Borneo Company but remained with the grantor of the lease to whose title the Philippines Government has succeeded. The present day Philippine claim to Sabah was first raised by Macapagal before he became President. In April 1950, he was co-sponsor of a resolution put before the House of Representatives calling for a rejection of the 'unwarranted and illegal' annexation of North Borneo. Although the resolution was passed it was then shelved because of the outbreak of the Korean War. Later, when Macapagal became President he raised the issue in earnest. Various appeals from Britain were dismissed and in June 1962 came the claim to sovereignty over the whole of Sabah. This was rejected and initially the Philippines refused to recognise the newly created Malaysia into which Sabah was incorporated. Not until 1966 were diplomatic relations restored.

It was not until the Sabah elections in 1967 that the issue loomed large again in South-East Asian politics. The elections were the first to be held in the territory following the ending of confrontation between Indonesia and Malaysia and were carried out, under United Nations supervision, to enable the local inhabitants to decide what form their political future should take. In consequence, the Philippines agreed to send observers but at the last minute President Marcos bowed to internal pressure and revoked his decision. There then followed a steady deterioration in relations between the two countries culminating in the breaking of diplomatic relations in 1968 (they were restored at the end of 1969).

The consequences of the present state of affairs are that they have held up two developments of major significance for the Philippines. For one, the recently signed anti-smuggling treaty with Malaysia is no longer effective with the result that the advantages gained from initial success in reducing the influx of cigarettes and other dutiable items into South-Western Mindanao from Sabah have been lost. The effects of the resumption of full scale smuggling can be gauged from the fact that some 6 billion cigarettes pass through Sabah annually en route to the Philippines; it is estimated that the duty lost on cigarettes and other smuggled goods has amounted to some £10 million since the Second World War. Secondly, the Sabah issue has put in jeopardy progress towards regional co-operation under the auspices of ASEAN (the Association of South-East Asian Nations). In view of the inherent instability of the area, this is an especially regrettable development. Fortunately, in late 1968 the Philippines and Malaysia recognised the important economic benefits of ASEAN. As a result there has been tacit acceptance by both sides that the Sabah issue is dormant, though far from dead.

Since 1969, therefore, the Sabah issue has been chiefly conspicuous by its absence, despite a complaint by Malaysia concerning an alleged violation of Sabah's air space by the Philippine air force coupled with accusations of build-up of armed forces in Southern Mindanao. Sabah did not appear as an issue in the 1969 Presidential Election presumably as President Marcos judged that any positive action would not bring any prestige. Consequently, as far as the Philippines is concerned, the issue is likely to remain dormant until the state of domestic politics makes a revival desirable.

References

General Note: Much detailed information in this chapter has been derived from unpublished data (including business intelligence reports) secured by the Economist Intelligence Unit to serve as the bases of its *Quarterly Economic Review* covering the Philippines and Taiwan. Thus it is not amenable to attribution.

1. Anon, *An Economic Profile: Philippines*, Sycip, Gorres, Velayo and Co., Makati, January 1969, p. 1.

2. R.E. Huke and J. Duncan, 'Spatial Aspects of HYV', *Proceedings of the Seminar on Economics of Rice Production in the Philippines*, International Rice Research Institute, Dec. 11-13 1969.
   The authors studied the IR-8 diffusion pattern in Nueva Ecija and recorded a rate of adoption exceeding that involving similar innovations in the United States and Western Europe. As of 1972 the rate of adoption of HYV rice by Filipino farmers was the highest of all developing nations.
3. A. Barrera, 'Classification and Utilisation of Some Philippine Soils', *The Journal of Tropical Geography*, Vol. 18, August 1964, p. 28. Data on which this article is based relates to approximately 65% of the land area but includes most of the more fertile areas. Considering the total land area of the Philippines, only one third probably is suitable for cropping.
4. F.L. Wernstedt and J.E. Spencer, *op.cit.*, p. 273.
5. *Ibid*, p. 260.
6. *Ibid*, p. 259.
7. A.R. Apacible, 'The Sugar Industry of the Philippines', *Philippine Geographical Journal*, Vol. 3 Nos. 3-4, July-Dec. 1964, p. 87.
8. J. Kirkup, *Filipinescas: Travels through the Philippine Islands*, Phoenix House, London, 1968, pp. 56-7.
9. In addition to attempting to cater for the more common needs of the public, the general store attempts to satisfy a miscellany of additional wants. Included in this category are: bicycle parts and accessories for those fortunate enough to be able to own independent means of transport; rice and maize husks for those enterprising individuals who attempt systematically to fatten pigs and chickens rather than let them run wild through the town scavenging where they may; and cheap suitcases for travellers.
10. Various surveys conducted by the Bureau of Census and Statistics as reported in Anon: 'Implications of Demographic Factors on Rural Development', *Journal of Philippine Statistics*, Vol. 19 No. 4, Oct-Dec. 1968, pp. XIII-XXI.
11. R.E. Huke, *op. cit.*, p. 197
12. D.J. Dwyer, 'Irrigation and Land Problems in the Central Plains of Luzon', *Geography*, Vol. 49, 1964, p. 238-46.
13. H. Von Oppenfeld, 'Some internal causes of rural poverty in the Philippines', *Malayan Economic Review*, Vol. 4, 1959, p. 42.
14. In Neuva Ecija a survey (R.E. Huke and J. Duncan, *op. cit.*, p. 30) has demonstrated that landlords will take the initiative, but only if share cropping is practised. In this way the landlord profits from improved yields, a state of affairs not possible when holdings are converted to fixed rent leasehold as a result of the implementation of the Reform Code. The leaseholder often is not able to employ HYV rice as he is unable to secure the necessary loans for fertiliser, pesticide, etc. unless he has a progressive landlord, for as yet the Reform Code (ACA) has inadequate funds for this purpose (see page 297). In other words the landlord's attitude is more critical than the nature of the tenancy.
15. H. Umehara, 'Socio Economic Structure of the Rural Philippines', *The Developing Economics*, Vol. 7, 1969, p. 321.
16. See reference 17 below.
17. The minimum economic size for an irrigated rice farm in Laguna Province has been calculated at two hectares, rising to five hectares for upland rice (J.T. Scott, Jnr. and P.C. Kuhonta.: 'Resources required for Viable Farms in Laguna Province, Philippines', *Illinois Agricultural Economics*, Vol. 8 No. 2, July 1968, p. 34).
18. N.W. Schul, 'A Philippine Sugar Cane Plantation: Land Tenure and Sugar Cane Production', *Economic Geography*, Vol. 43, 1967, p. 169.

19. F.L. Wernstedt and J.E. Spencer, *op.cit.*, p. 191.
20. F.L. Wernstedt and P.D. Simkins, 'Migrations and the Settlement of Mindanao', *Journal of Asian Studies*, Vol. 25, 1965, p. 90.
21. K. Hausherr, 'Agricultural Colonisation in the Kapatagan Basin, Lanao del Norte, Mindanao, Philippines' (in U. Schweinfurth (ed.), *Problems of Land Use in South Asia: Yearbook of the South Asia Institute, Heidelberg University 1968/69*, Wiesbaden, 1969), pp. 100-16.
22. F.L. Wernstedt and P.D. Simkins, *op.cit.*, pp. 95-8.
23. D.J. Dwyer, 'The Problem of In Migration and Squatter Settlement in Asian Cities: Two Case Studies, Manila and Victoria Kowloon', *Asian Studies*, Vol. 2 No. 2, 1964, p. 155.
24. T.G. McGee, *op. cit.*, pp. 141-5.

9
The Future

In the following pages some indication is given as to how present patterns of life and livelihood in the Philippines may alter in the coming years. Certain developments can be forecast with some accuracy. In the field of economic geography, for example, an upsurge in copra production as newly planted coconut trees reach maturity is probable. Likewise, in the field of social geography an increase in the influence of the urban (or rather suburban) middle class can be forecast with some confidence. However, even such predictions can be upset easily in a country such as the Philippines where political and financial instability vie with climatic and tectonic instability in inserting a significant element of uncertainty into the evolutionary pattern of the life and livelihood of the average inhabitant.

For the foreseeable future the role of the cities can only increase even though, as McGee[1] rightly points out, the current growth of South-East Asian cities is not necessarily soundly based, thus leading to possible long term instability. Nevertheless, the approach in this chapter is similar to that of Chapter Eight, in that rural matters are given prior treatment when considering the pattern of social geography. A cursory glance through the following pages will reveal much cause for optimism. But the most optimistic progress anticipated, however significant its impact on the pattern of economic and social geography, must not disguise the fact that the Philippines is far from graduating from a 'developing' to a 'developed' nation. This is a topic which has been widely discussed particularly in the field of developmental economics; in this particular instance it will suffice to point to the profound contrast with the Philippines' near neighbour to the north—Japan:

'The flight from Tokyo to Manila is not simply from a Haneda Airport, served by a monorail, equipped with all the latest gadgetry, and superbly provided with numerous comforts, to a Manila International Airport, served by largely poor access roads, an airport once 'advanced' but now according to the pilots who use it, inadequately provided with essential mechanical aids. It is, also, a flight across the world's major international boundary ... between those who are developing the capacity to improve their circumstances and those who are still largely at the mercy of circumstances they cannot, or will not control; between those areas where poverty is diminishing and those where poverty is increasing; between a country which has brought population growth under control and one which has not even produced accurate statistics on its very rapid population growth ...

'You pass from Japan where economic activity is *the* status activity, to the Philippines where politics, while it may have lost the status of respect, certainly has the capacity to attract a disproportionate investment of the nation's energy. In any case Japan progresses at ever-increasing speeds despite politicians who appear as marginal to Japanese progress as they seem central and crucial to getting 'progress' started in the Philippines. The Filipino politician uses the nation as a stage, and is not above inducing violence to keep himself upon it. Accommodation is usually for personal ends, compromise in the national interest conspicuous by its absence.

'In Tokyo, even the lights of government offices burn late at night, and not just for the cleaners. Officials may be working without overtime pay because they feel a personal obligation to the job, or to a senior colleague. In Manila it is not difficult to find people asleep over their desks during office hours. Personal loyalties motivate many things, but seldom, it would seem, unremitting hard work.

'From a people bent on organising themselves towards a higher standard of living, to a people much more richly endowed by nature, but not by themselves, is some transition. But then, the Japan one leaves behind has no doubts about its national identity, does not require of its politicians that they create national unity, and (at present) finds no need for a high degree of national assertiveness in order to conceal a lack of national self-confidence, purpose, and achievement.'[2]

A. Economic Geography

Indicators of Philippine economic growth (Fig. 44) suggest that no dramatic acceleration or deceleration is anticipated in the 1970s (the revised Second Four Year Plan, covering the period 1972 to 1975, is equally conservative in outlook suggesting that only the mining sector will exhibit an exceptional rate of growth). Prospects for the Philippines contrast with those predicted for Japan which is expected to exhibit a slow down in economic growth in the 1970s. But such a comparison needs to be put in perspective, for whereas Philippine growth, measured in real (constant prices) GNP, will be rising annually by about 6%, on average that for Japan will be a far more impressive 10%. Furthermore, whereas the growth rate for Japan's population, 1% per annum, will be less in the 1970s than in the 1960s, that for the Philippines will be about one tenth greater than the already exceptionally high rate (approaching 3.5%) experienced in the 1960s. If the comparison with Japan is felt to be unwarranted because of the nations' greatly differing stages of development, a parallel state of affairs exists with respect to a more similar, and closer, neighbour—Taiwan.

Taking the broad perspective, therefore, prospects for the 1970s are not encouraging. In the domestic context, food production is the key. The planners still confidently predict that self-sufficiency in cereals and protein is imminent but a more realistic appraisal suggests that a measure of self-sufficiency sufficient to cover periodic short-falls in supply due to inclement weather and marketing inadequacies, not to mention rising per caput consumption, still is some way from being attained on an overall basis. The necessary imports to meet consumers' food needs (not to mention all the other foreign consumer products to which the Filipino is addicted) will continue to require that export expansion be given top priority. This, however, will not be easy to secure at an adequate rate for in the 1970s Philippine exporters will face not only the universal difficulties that beset primary producers in the 1960s but also a possible contraction in world trade and, of course, the realignments resulting from the ending of the Laurel-Langley Agreement. International action such as that stemming from UNCTAD 3 (the third meeting in Chile of the

UN Commission on Trade and Development) may bring some encouragement but, whatever the outcome, the Philippines will need to intensify current efforts to maximise the added value of

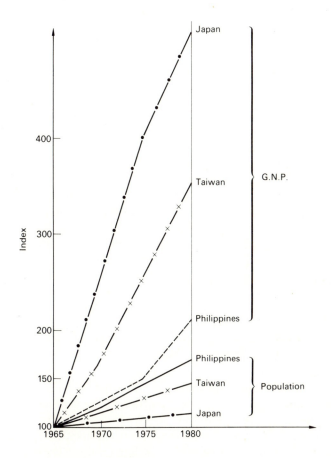

Fig. 44. Comparative Estimates of Population and G.N.P. Growth for Japan, Taiwan and The Philippines, 1965-80

export commodities and to encourage import substitution if the country's external trade is to be balanced.

The above picture of Philippine economic growth in the

1970s could of course be modified significantly in practice. The imminent ending of the Laurel-Langley Agreement in its present form is bound to alter the country's external trade in both direction and content. Even more apposite has been the drastic devaluation of the *peso*; this development, coming right at the commencement of the decade, means that many predictions may need substantial revisions. Further, more speculative factors which might complicate the situation in the 1970s include a possible breakthrough in land reform and an increased government preoccupation with social goals leading to a substantial increase in expenditure allocated to services for minimal or low tax payers.

In terms of economic geography, realistic predictions suggest that in general changes will be of detail rather than of substance. But some are likely to have significant regional impact (Fig. 45). Outstanding in this respect can be included such developments as the ever increasing dominance of Manila; iron-nickel mining at Surigao in Eastern Mindanao; the Pan-Philippine Highway; the universal utilisation of HYV rice, but particularly in relation to the completion of the vast Pampanga River irrigation project in the Central Plain of Luzon; realignment of the sugar industry in the post Laurel-Langley era; and the expansion of copra production in Mindanao. In addition, there are further substantial projects now in their initial stages which may make an impact in the latter part of the decade. These include the conversion of 30 000 hectares of the Candaba swamp on the Central Plain of Luzon into the largest area of freshwater fishponds in Asia; the establishment of the country's second largest electric power using unit—a 60 000 ton capacity aluminium smelter to be located in the industrial complex of the Misamis Coastal Belt; the proposal to increase commercial banana production in Eastern Mindanao sevenfold for sale on the rapidly expanding Japanese market; the Mariveles free trade zone; and the five year programme to produce 5 million sacks of maize for export on a cooperative basis in the vicinity of Tuguegarao in the Cagayan Valley. Finally, in a negative vein, there is a possibility that the unprofitable gold mining industry of the Baguio area will close down with government support being directed solely to mineral deposits elsewhere from which gold is secured as a by-product.

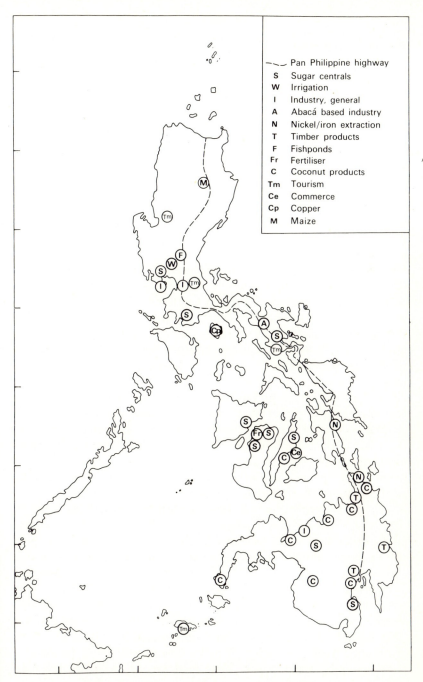

Fig. 45. Selected Development Nodes (Recent or Planned)

Primary Production

In this sphere progress on all fronts can be expected. Agriculture will continue to lead the way but will perhaps decline somewhat in relative importance as a result of large scale forestry developments coupled with the exploitation of newly discovered mineral resources and, perhaps, further successful prospectings. However, progress is expected to be inadequate to keep pace with population growth where food resources are concerned. One particular facet—meat production—forms the subject of a special case study below. The disappointing past performance of cash crops may well be repeated if current pessimistic forecasts of world trade levels are confirmed; if producers cannot adjust to the ending or modification of the Laurel-Langley Agreement; if competition from other primary producers and synthetic substitutes cannot be met; and if soil erosion is not checked. On the other hand, given no more than its fair share of typhoons and other expected natural disasters, the progress made in the fields of mechanisation, fertiliser application, farmer education, product processing and export marketing provides a base on which relatively efficient cash crop production can flourish.

Agricultural development therefore needs to be centred on improvements in productivity. In the post-war years agricultural growth was rapid for it was operating from a base depressed by the consequence of Japanese occupation. Furthermore, by the imposition of forced savings on the agricultural sector—particularly the export sector—industrialisation was subsidised at the expense of the real welfare of the farm worker.[3] This policy proved tolerable because agricultural growth exceeded that of agricultural population but, with limited scope for new land settlement, real welfare of farm workers can only be sustained, let alone improved, by means of greater efficiency in agriculture.

Improvements in productivity include a realignment of government fiscal policy. Rice self-sufficiency, for example, should make redundant present rice price subsidies worth over £10 million per annum. Such funds would best be switched from encouraging production to improving processing facilities, for example. Even more beneficial would be a switch of emphasis to boost feed grain production as a means of satisfying rising demand for meat, poultry and dairy products. Finally, government support is needed urgently to extend the current agricultural loan services provided to embrace medium and long term loans. Tree crops, notably coconuts, would benefit especially from such a move, as would agricultural processing in general.

Dramatic modifications of the land use pattern are not expected in the 1970s. Change in a rural environment, particularly one such as the Philippines where tradition rather than profit is often the dominant influence, is rarely swift but even where advances in technology and adequate capital are available, as will be shown below may be the case with abacá, any change is essentially marginal in nature. It requires a profound modification of existing production or marketing conditions to cause, say, the disappearance of the tobacco lands of the Ilocos or the transformation of the sugar lands of Negros into rice lands. The latter has been mooted in some quarters in view of the likely adverse effects of the ending of the Laurel-Langley Agreement but does not take into account the high degree of land use inertia resulting from the substantial level of long term capital investment. In the second half of the 1970s the sugar lands might begin to contract at the margin, where least favourable net returns are experienced, but in the core areas the intrinsic suitability of the land for sugar almost certainly will make possible net returns sufficient to encourage producers at least to see out the life of their existing capital investment.

Only in one sphere of land use are dramatic modifications expected, or rather feared. This relates to the forestlands where deforestation is a cause of serious concern both from the unscrupulous activities of commercial loggers, and the increasing rate of *kaingin* and allied squatting resulting from growing population pressure. Recent aerial photography has confirmed that reckless timber cutting has deprived Mindanao's future forest potential of 1 million hectares in the last decade alone. Over the country as a whole 1·4 million hectares are in urgent need of afforestation, i.e., critical water-shed areas. But funds are made available at such a level that only 25 000 hectares can be dealt with annually. The consequences already are evident in the upper Agno River where the life of Ambuklao Dam has been halved because of the accumulation of sediment washed down from its watershed; likewise, a thirty ton per day pulp mill in Bataan has been forced to close down because its wood supply had been exhausted through unwise exploitation. Some forest operators have a more realistic approach but many 'cut and run' loggers remain, not to mention the *kaingineros* and other small scale violators of forest laws. With only one guard per 17 000 hectares of public forest the outlook is bleak indeed.

In contrast to land use, the 1970s could see major changes in the pattern of life and livelihood in certain rural areas; for example, the spread of coconut into Mindanao or the withdrawal of abacá from Leyte. Such change in land use at the margin may reflect profound structural change in the core area. Change, whether it be caused by inability to master physical conditions, such as typhoons and *cadang cadang* in the coconut lands of Bicol; radically changed markets, such as might be the case with sugar in the late 1970s; or technological change, such as HYV rice, is soon felt by the man in the street. Any adverse implications are the more speedily felt also, as the imperfections of Philippine society often hamper dissemination of the benefits of progress.

To illustrate the impact of change in rural areas one can do no better than refer the reader to a recent assessment of future Philippine rice requirements.[4] It has been determined that an increase in rice production of about 200% will be necessary in the period 1965-90. By 1990, if all land still available for cultivation is developed, 9.6 million hectares (exclusive of area in second crops) should be available, representing an additional 2.1 million hectares over 1965. With 3 million hectares needed for rice in 1990, only 0.9 million hectares, or 43%, of the additional land can be used for crops other than rice. But the above calculations make one important assumption—that adequate irrigation water is available so as to utilise the potential of HYV rice. Given adequate appreciation of the need and the resultant substantial funds, the Philippines has the water resource capacity to extend the 1965 irrigated area by almost 2 million hectares by means of gravity and pump schemes plus a further 0.4 million hectares if storage irrigation schemes are developed also (these extensions represent increases of 267% and 53% over the estimated 1965 irrigated area of 0.75 million hectares). Furthermore, massive increases in other inputs besides irrigation water will be necessary; for example, fertiliser with a nitrogen equivalent of up to 284 000 tons would be required by 1990, or no less than 19 times the 1965 quantity.

In practice it is doubtful if the optimum development outlined above will materialise. Nevertheless, the trend of rice development has been established and to a greater or lesser extent the rice lands of the Philippines will be influenced by it in the coming years. The resultant modification can, and should, have profound implications for the life and livelihood of many Filipinos.

Before considering what the 1970s hold for the land and people involved in meat and abacá production, whose current features were studied in some detail in the previous chapter, it is advisable to review the future of a commodity vital to agricultural production in the 1970s—fertiliser. In the previous chapter the review of soil erosion in the Philippines pointed up the totally inadequate usage of fertiliser in the 1960s. It will be demonstrated that a significant change in attitude to fertiliser will occur in the 1970s reflecting a substantial increase in the overall efficiency of Philippine agriculture.

Historically, the sugar industry has been by far and away the leading consumer of fertiliser, accounting for almost half of nitrogenous fertiliser in 1967 and well over half of all other types (Table 33). With sugar accounting for only about 3% of the crop area, per hectare consumption has been minimal outside the sugar lands but a dramatic change is expected in the 1970s. The changing pattern of fertiliser consumption is expected to result principally from even more widespread use, and of greater importance more efficient use, of HYV rice and maize whose success in the latter half of the 1960s has been shown to be partially dependent on adequate application of fertiliser. Given also a general awakening in interest in fertiliser on the part of Philippine farmers as a result of farm extension work by the government and through promotion by the fertiliser producers, it is anticipated that the average growth rate for fertiliser usage in the period 1967-80 will be 8·6% per annum. Even if this proves to be too optimistic a forecast there is little doubt that fertiliser consumption will increase at a much faster rate than food production, which is predicted to grow at 3·5% per annum.

Of equal significance will be the changed nature of fertiliser end usage. The share of sugar in total fertiliser consumption will

TABLE 33
Fertiliser Consumption 1967-80 (tons)[a]

		Sugar	Rice	Maize	Coconut	Others	Total	Average Annual Growth Rate (%)[b]
	Total	67 131	34 421	4163	2675	21 989	130 379	—
1967	Nitrogenous	37 829	25 528	3396	1380	11 348	79 481	—
	Phosphatic	17 033	6722	515	803	4584	29 657	—
	Total	120 691	133 169	68 656	22 132	34 907	379 555	8.6
1980	Nitrogenous	53 001	81 700	45 680	9217	19 696	209 294	7.7
	Phosphatic	26 590	41 950	22 724	9991	7209	108 464	10.5

Source: Economist Intelligence Unit.

[a] In terms of nutrient content only—nitrogen content of nitrogenous fertiliser is only one fifth by weight; superphosphate is only one third phosphate.

[b] Compound from 1967.

fall sharply in the future, though increasing in absolute terms. By 1980 rice and maize, which together occupy some ten times the cultivated area occupied by sugar, will together account for over 56% of total fertiliser demand (compared with less than 30% in 1971). This will reflect average annual growth rate of 10·5% and 24·0% respectively. Consumption by coconut producers will increase nearly ninefold (or 17·6% per annum).

To meet the anticipated demand will entail a major expansion of fertiliser output. Fortunately, the four existing producers—located at Toledo (Cebu), Taguig (Rizal), Limay (Bataan), and Iligan City (Lanao del Norte)—currently are operating well below capacity (only 33% in 1971). Furthermore, plans are in hand to expand the Iligan City facilities which currently only manufacture nitrogenous fertiliser. Likewise, the Negros based Sugar Producers Marketing Association has plans to establish production facilities to meet local needs.

Meat. It appears quite probable that within a few years livestock production will emerge as a significant element of the rural scene. Development in this sphere of agriculture has the powerful stimulus of a demand that greatly exceeds supply. The resultant efforts of free enterprise to expand production can be expected to be reinforced by government action stimulated both by a desire to reduce the level of imports and the need to improve the protein element of the nation's diet. Recent forecasts indicate, perhaps optimistically, that self-sufficiency in pork and poultry meat could be achieved in 1973, and self-sufficiency in beef could be achieved by 1975. Hopes for attaining the poultry meat target have been enhanced with the 1971 announcement of a £1·75 million project for South-Western Luzon to produce 3500 tons annually.

Government action to stimulate meat production is taking a variety of forms. Financial aid is provided by the special loan programme for cattle and pig producers inaugurated by the Development Bank in 1967. For each year up to 1980 funds will be made available to cattle ranchers, both large and small, to finance a variety of improvements and the purchasing of breeding stock. The aim is to increase output by half to exceed 2 million carcases per annum. In like manner, pig producers are being encouraged to double registered pork production by

1980. Upgrading of the quality of research and instruction in animal husbandry at the University of the Philippines is planned to help ease the shortage of skilled management. At the other end of the scale, meat marketing in Manila should show a marked increase in efficiency with the completion of a new terminal market with its cold storage facilities. Overall economic development will also benefit livestock production. Achievement of self-sufficiency in rice and maize should ease the shortage of animal feedingstuffs, as will improvements in grain drying and storage. Improvements in secondary roads, too, will help to ease distribution bottlenecks.

The awareness by private enterprise of the prospects for livestock production is highlighted by the recent formation of an association of cattle raisers actively canvassing for improvements in beef production. On the farms themselves major expansions in feedmixing are anticipated and crossbreeding of the small native cattle with imported stock is gaining momentum. Such independent action may well prove to be of more significance than government sponsored schemes whose policies and finance are subject to unpredictable vacillations.

By the 1980s in selected areas of Luzon and Mindanao livestock production should dominate the rural scene both directly, and also indirectly, by stimulating irrigation development and the more efficient production of rice, maize and other feedingstuffs. The expansion of pig numbers in the form of the larger production units may lead to their becoming less urban orientated as distribution bottlenecks are eased; as public health measures are tightened; and, most important, as the need to be located close to feed supplies becomes essential in the face of ever growing scales of production.

Increased supplies of red meat at reasonable prices will improve the quality of the Filipino's diet with all its attendant consequences. In the long run systematic livestock production will cause a decline in small scale backyard production; *carabao* meat production also will decline as a result of the availability of better quality meat and farm mechanisation. Both the urban and rural scene thus will change. The consequences of these developments will only be apparent in the long run but, even in the short term, livestock production, marketing and consumption will undergo significant changes leading to an upgrading of its importance to the economic geography of the Philippines.

Abacá. The future for abacá appears bleak (the 1971-4 Development Plan projected an annual growth in production of only 1·4%) but there is some cause for cautious optimism nevertheless. The reduced importance of abacá as an element of the rural scene is likely to persist despite the efforts of the Abacá and Other Fibres Development Board. Resolving the industry's problems involves the establishment of a realistic level of production in the best suited locations. No prediction of production levels can be made at this juncture but it is likely that a further contraction of the producing areas could take place for, even if efforts to reduce production costs are successful, substantial end uses other than cordage need to be found. Abacá can be expected to find wider usage for such products as rugs, mats and sacks but consumption levels so attained are not likely to be substantial. It is demand for abacá as pulp for paper making that offers the best prospects, but only if substantial problems related to capital investment and modified production and harvesting techniques can be overcome. A further complication is the fact that in this sector, too, abacá could face increased competition from substitutes.

From a geographical point of view the implications of the direct use of abacá in paper making are considerable. In comparison with wood, abacá has such a small fibre content that large areas of land, large quantities of labour and considerable energy inputs are required for economic production. The implications of the establishment of milling facilities thus would be considerable. At Canlubang, in the province of Laguna, abacá is processed into pulp in conjunction with bagasse, while at an Iligan City factory abacá and timber form the raw material for pulp production. The factory to be established in Camarines Sur, however, will be employed to process abacá exclusively.

The lower grade of abacá required by the pulping process might be thought to be to the advantage of small scale producers but this is unlikely to be the case. With continuity of supply essential, large supply contracts can be expected to be the order of the day. Small scale producers, excluded from supplying the mills and also excluded from the cordage market by their above average production costs and their lower quality product, would appear to have a limited future, their only opportunity being to supply the less demanding local market.

The abacá industry thus can look forward to two distinct

futures. The first, and more probable, will see a further decline in production, as marginal producers are eliminated, to reach a stage whereby abacá forms a prominent part of the rural scene only in selected parts of Bicol and Mindanao. Alternatively, successful processing of abacá for pulp will give the industry a new lease of life, based predominantly on large scale production. Whatever the outcome, a focusing of activity upon the best endowed production areas appears inevitable.

Manufacturing

Manufacturing in the 1970s in the Philippines is unlikely to increase substantially its contribution to NDP. However, it will be the sector most likely to benefit from redeployment of work force: by 1974 it is predicted[5] to be employing 12·8% of the work force compared with 11·9% in 1969 (c.f. agriculture, forestry, fishing, etc., down to 55·0% from 56·4%). Government policy is designed to increase its contribution but the pronouncements of politicians and financial pundits of a dramatic growth in industrialisation leading to a subsequent stabilising of the Philippine economy would not appear to be based on fact. In certain sectors such as timber, abacá, mining, iron and steel, sugar and even aluminium there are grounds for optimism, some of it based on current achievements, but it is an inescapable fact that the current restraints to industrial growth are unlikely to disappear in the 1970s.

Formidable barriers to development exist. Perhaps outstanding is fragmentation of effort resulting from the inability of the government to control current excesses of *laissez faire* economics; for example, in the cement industry in 1968 the twelve existing producers were adequately able to satisfy demand but, with ten more plants planned for the immediate future and a further forty under consideration it has been estimated that, despite the strong growth in demand, by as soon as the early 1970s production capacity will be at least 50% greater than demand. In addition, there are the problems caused by the shortage of capital resulting from, *inter alia*, the instability of the economy and a policy of economic nationalism that has deterred foreign investors; the predicted absence of any substantial growth in consumer purchasing power; and the under-utilisation of existing facilities. The latter result in part from fragmentation and in part from over optimistic forecasts of demand at the time of factory construction—the country's two ammonium sulphate producers in 1967 operated at only 11% of capacity and, even after allowing for growth in demand, probably will be operating only at 43% of capacity in 1974.

From a geographical standpoint the absence of any substantial relative growth in manufacturing will mean no major changes in location patterns. In particular, the dominance of the Manila metropolitan area will not be challenged. This despite the fact that few of the more spectacular new developments such as paper production from abacá or aluminium smelting will be located there, and despite efforts to relocate export industries, notably embroidery, in Bataan at the Mariveles free trade zone (plans here call for a diverse export output approaching £100 million and employing 40 000 persons by the mid 1970s). The metropolis contains those industries basic to the economy—food, beverages, textiles, chemicals, etc.—that can be expected to record steady growth in the 1970s. In this context the chief challenge would appear to spring from the inter-regional centres, with Cebu outstanding. These increasingly should attract branch consumer goods plants to serve more economically the growing population of their hinterlands. Manila's chief rival as an industrial centre will continue to be the Misamis Coastal Belt. Here, continued expansion of iron and steel, fertiliser and associated industries could be given a substantial boost if an aluminium smelter is established there as planned for the late 1970s.

Perhaps the most significant contribution to the changing pattern of economic geography will result from dispersed manufacturing development associated with the country's natural resources, and sugar, abacá, minerals and timber in particular. Individually a new sugar mill, an abacá based pulp and paper mill, a nickel smelter or a plywood and veneer factory, however large, make but a small contribution to the national scene. But to the inhabitants of the area in which a factory is located, and for the people attracted to the area by its development, its establishment means a transformation in both life and livelihood.

Ultimately, one could expect to see the present concentrations of activity in Manila and the Misamis Coastal Belt being joined by sizeable secondary centres of manufacturing activity in Cebu, Tarlac, Bacolod, Iloilo, Davao and similar urban foci of industrial development and complemented by a wide range of rural development nodes. In the foreseeable future, industrial growth will be restricted to the most favourable locations; small scale cottage or experimental manufacturing facilities may well be found elsewhere but their impact on local life and livelihood will be marginal.

Again, if the government's industrialisation policies were to bear fruit, manufacturing development would be diverse in terms of not only location but also materials handled and, more important, the end product. But efforts to diversify production and increase its value added content face not only the obstacles mentioned above but also those common to most developing nations. These include a lack of skilled labour and management, poor transportation facilities, inadequately surveyed natural resources, restricted local markets and high production costs which generally preclude export market penetration.

Processing and assembly operations will thus remain the cornerstone of Philippine manufacturing in the 1970s. Such operations, therefore, form the subject of the following case-studies which are concerned with copra crushing and motor vehicle assembly. The predicted development of these spheres of action could well illustrate the way in which much of Philippine manufacturing will evolve in the 1970s.

Copra Crushing. The ending of preferential trading agreements with the United States means that the coconut oil quota will be abolished in 1974. This will parallel a reorientation of Philippine exports of coconut products in general in the face of the expected severe competition from lower priced sources such as Ceylon, and possibly from synthetic substitutes. This will involve a search for new markets or greater penetration of alternative outlets notably those in Europe, coupled with increased marketing efficiency to retain as much as possible of existing trade with the United States. Whatever the outcome a continuation of the trend towards increasing the value of products exported by prior processing can be expected, of which the crushing of copra to produce coconut oil is a prime example.

Distribution is the second major problem requiring solution

by the crushers in the 1970s. Currently, costs of transporting copra to the crushing plants are high due in no small measure to the poor roads in producing areas, inadequate seaborne traffic, and the location of most mills in Manila far from the origin of the copra. With Mindanao emerging as the centre of coconut production, the copra crushers also will need to respond to the consequent shift in location and associated changed pattern of marketing which will result. Even if distribution bottlenecks are eased by the implementation of the government's infrastructure planning, a tendency for new plant to be located at or near the production areas can be expected, if only to minimise copra deterioration en route. In the long term most large coconut producing areas, especially Mindanao, can expect the introduction of modern processing facilities with resultant beneficial effects on local employment. If the efforts of the newly formed Coconut Research Institute are successful, the establishment of such plant could also foster the growth of ancillary industry using by-products such as coir and coconut husk.

The third restraint to growth is the most difficult to overcome. A greater emphasis on copra crushing, and efficient crushing at that, will mean that existing mills will require drastic modernisation for most are furnished with obsolete equipment. This will require substantial capital investment as will the new plant required to meet the expected expansion of demand for coconut oil whether for export or for the home market. If the bulk of copra produced is to be crushed by local facilities it is estimated that coconut production could support up to twenty additional mills each meeting the minimum requirements for an economic operation of 18 000 tons per annum.

Mindanao obviously stands to benefit most from any expansion and relocation of copra crushing facilities. However, it should be stressed that firm plans involving major facilities announced by the copra crushers currently have been restricted to a single plant, to be located in Surigao. The above mentioned problems facing the industry obviously are a powerful deterrent, but by the mid 1970s, with the Laurel-Langley Agreement at an end and the new plantations in Mindanao maturing, the copra crushers will have to face up to the future. Then, major agricultural service centres in Mindanao such as Butuan, Davao, Ozamis, Cagayan de Oro, Cotabato and Zamboanga City could well be chosen as the sites for new large scale mills. Success in the operation of these mills needs to take into account also the Mindanao climate with its regular rainfall. Copra drying will be less able to rely on the sun as is the case in traditional locales further north so it will be essential that modern kiln drying facilities be provided, for a reliance on the traditional *tapahan* system will seriously affect copra quality.

Once the government's coconut replanting loan programme is completed funds may be made available for kiln drying and even copra crushing facilities. Private investment probably will predominate however, although in the case of copra crushing its large scale may involve foreign investment. This may not be easy to obtain in view of the government's attitude that foreign investment in local business, unless closely restricted, is detrimental to the national interest. Considerable incentives may be required to enable the industry to secure the capital it needs; perhaps the impending re-orientation of the copra trade will provide this, possibly in the form of American aid. Until then the industry epitomises the dilemma faced by a developing nation—first, whether to allocate scarce resources into a specific industry so as to improve export quality and boost foreign exchange earnings, or else risk the consequences of low quality exports in an effort to build up manufacturing as a whole; and secondly, whether to encourage overseas investment to secure more rapidly economic benefits or to sacrifice economic development for economic independence by discouraging it.

Motor Vehicle Assembly. Industry sources expect that the market for motor vehicles in the Philippines should reach 50 000 units in the middle or late 1970s. The motor vehicle assemblers thus will need to more than double output in this period. Furthermore, the growth trend could be an underestimate for the government's current road building programme, so far successfully implemented, should stimulate motor transport further. In particular, the completion of the 1300 km Pan-Philippine Highway could reduce the importance of water borne traffic. Less spectacular but equally significant are the general improvements to existing roads and bridges coupled with the provision of additional feeder roads; these are doing much to facilitate the movement of goods and passengers by land which until now has been at best uncomfortable and, at worst, hazardous in many parts of the country. Again, the market could grow even more rapidly if the tax burden were eased so that costs and consumer prices could be reduced; a reduction of credit, too, would further stimulate sales. The market for locally assembled vehicles also could be significantly increased if the government could effectively discourage the direct importation of built-up units (the government itself makes most of its major purchases from overseas suppliers in order to avoid payment of the varied taxes that it imposes upon the industry).

The 1970s, therefore, could well be a boom period for the motor vehicle assemblers even if overall economic growth is slower than forecast. For motor vehicle manufacture (as opposed to assembly) to become practical in the Philippines, annual demand should exceed 30 000 units. With this level of demand rapidly approaching, the industry must come to grips with the current fragmentation of supply sources. At present, practically all the main automobile manufacturing countries have their major makes represented in the local market. The profusion of makes and models tends to increase assembly costs, as many of the assembly plants are working at quite low volumes, and distribution costs are also higher. Apart from higher advertising and selling costs, there is the high cost of maintaining spare parts inventories. It has been argued that government intervention may be needed to bring about a reduction in the number of local suppliers, perhaps coupled with a revision of import control measures to ensure that the consumer's choice is not over-restricted. However, initially the industry will probably begin to rationalise as competitive pressures push profit margins down to a point where low volume producers and distributors simply will be unable to operate profitably. Then, the number of producing plants will be reduced and the surviving models will be able to build up volume, assisted by the normal growth of the market, perhaps coupled with price reductions made possible by tax incentives designed to encourage local manufacturing. The stage will then have been set for the industry progressively to manufacture more and more of the parts needed to assemble finished vehicles. Both export orientated and domestic production can be expected to be located in or near the Manila metropolitan area in the foreseeable future constituting as it does the principal port of entry for motor vehicle parts, the chief source of supply of skilled and semi-skilled labour and the dominant domestic market.

Already there are signs that a change in the industry's structure is on the way. The Ford Motor Company has taken the lead with its announcement of plans for a £40 million motor body and chassis plant in the Mariveles free trade zone as part of its 'Asian car' strategy. Subsequently, Toyota of Japan claimed to be planning a passenger assembly line capable of handling 700 vehicles per month. Chrysler, Volkswagen, and Renault also are considering the establishment of major assembly facilities. The most recent development involves a planned £11 million transmission plant to be built by General Motors, probably at Marieveles but possibly on the Misamis Coastal Belt. Both Ford and General Motors seem assured of Board of Investments support and at least two other manufacturers also will be allowed to participate in the government's plans for motor vehicle manufacture (Chrysler, Toyota and Volkswagen are expected to be the ones favoured).

Transport and Communications

As befits a country characterised by extreme physical fragmentation, resulting both from sea and land barriers, the government's Second Four Year Plan continues to place great emphasis on investment in transport facilities (41% of the proposed infrastructure investment, for example, is earmarked for highways). As in the previous Plan period ending in 1970 the railways are to benefit from improvements to rolling stock, repair shops, and signal and telecommunication systems; improvements planned for seaports include the construction of additional berths; and the upgrading of three-quarters of the nation's landing fields is to continue.

Air transport has been the subject of special attention on account of its superior mobility and flexibility: aviation can do most both to disseminate the ideas and decisions emanating from Manila (not to mention the steadily increasing number of tourists) and also to bring into relatively close contact the scattered segments of the nation. The traditional means of achieving these aims—inter-island shipping—continues to have an important role but, except where bulk commodity marketing is concerned, it increasingly will serve only as a supplement to both air and land transport. Less attention is paid to communications in current government planning as this has been the subject of past action. Internal telecommunications were accorded priority as they provide the simplest and speediest means of inter-island contact. Although the overall system still leaves much to be desired the recent completion of the long distance telephone network has finalised the basic pattern of infrastructure.

It was realised that the above developments would bring no final solution to the problems of this sector of the economy. Consequently, a United Nations sponsored survey of all transport modes was carried out with a view to formulating a systematic development plan for the 1970s. By determining investment priorities, establishing feasibility studies for high priority projects, and preparing recommendations for strengthening existing transport planning organisations and for increasing the efficiency of regulating and improving transport services, it is hoped that Philippine transport in the 1970s at last will be noted for its quality and efficiency. If so, all sectors of the economy and society in general will benefit. The consequences for the road system of improvements at present being implemented and planned, set out below, serve to illustrate clearly the gains that could be attained in the 1970s.

Road Transport. Improvements to the road system are vital to the country's economic development. This is evident especially in the remoter parts of the archipelago where the construction of a road of any class is an event of major significance. It either obviates the need to rely on more primitive modes of transport such as the *carabao*; or ensures more regular vehicular service which, for example, formerly utilised a river bed in the dry season; or it encourages settlement: in parts of Mindanao settlers took up occupance as soon as the right of way for a highway was established.

The need for a better road system is fully realised by the government. At present the emphasis is on improving the existing system and the provision of new feeder roads; following the success achieved in the First Plan, the revised Second Plan (1971–4) envisages an improvement and extension programme involving 20 000 kilometres of roads and 15 000 lineal metres of bridges. Later in the 1970s the recently completed United Nations transport survey is expected to serve as the basis for extensions to the main highway system, particularly those branching from the Pan-Philippine Highway.

Whatever the developments initiated it is probable that the increase in vehicle ownership will outpace the capacity of the road system. Despite the financial barriers to vehicle ownership, the number of vehicles registered in the Philippines is rising rapidly and any increase in the efficiency of the system will only stimulate it further. A quadrupling of vehicles registered by the end of the decade appears likely whatever the pace of economic growth in the 1970s.

Significant changes in the nature of the vehicles on Philippine roads also can be expected. The current increased popularity of small cars can be expected to continue as, leaving aside the cost advantage, it reflects improvements to road surfaces, and also the increased difficulty of driving a large car in congested Manila. Better roads should see jeeps decline in popularity but the multi-purpose *jeepneys* will remain popular as long as intense competition persists in the public transport sector.

Changes in the type of vehicle seen on the country's roads will be the most obvious, but very far from being the only, manifestation of the contribution a more efficient road system will make to the nation's overall economic development. The gradual elimination or minimisation of distribution bottlenecks such as between the Central Plain of Luzon and the Cagayan Valley; the improvement of access to the Bontoc and Bicol Peninsulas; the establishment of all weather roads in Central Mindanao; and the provision of a comprehensive urban motorway system in Manila can be expected to have major economic and social implications. Likewise, less spectacular developments such as the elimination of the east-west distribution bottleneck on Leyte between Tacloban and Baybay; the improvement of access to the Banaue terraces in Mountain Province; the establishment of all weather roads in the Hilongos, Leyte, hinterland; and the provision of paved roads throughout Hilongos municipality itself will contribute much to the well-being of the local inhabitants even if the national gain is slight.

Improvements to the road system can be expected to stimulate new activities—tourism is an obvious example. The growing number of car owning *Manileños* doubtless will welcome the opportunity to drive in comfort to view the Banaue terraces and Mt. Mayon. So too will the ever growing number of overseas visitors, predicted to rise from 73 800 in 1965 (and 144 431 in 1971 who spent some £13 million) to some 600 000 in 1980,[6] who will expect to find good roads on disembarkation whether it be Manila or an island in the Sulu Sea.

There also are likely to be intangible benefits such as the speedier conveyance of mail and the increased mobility of the rural population. These contribute to both economic and social development but the end result is the same—a further strengthening of the Filipino's control over his environment.

Commerce

In the 1970s the pattern of Philippine international trade could well undergo a modification as drastic as that induced by the Spanish. They substituted a long distance trans-Pacific trade pattern for the traditional short hauls south and west to what is now mainly Indonesia and, in doing so, moved the centre of commercial gravity of the Philippines from Mindanao to Luzon. Already the trans-Pacific element of Philippine international trade has been weakened by the resurgence of Japan. The ending of the Laurel-Langley Agreement, the growth of ASEAN, and Australia's increased involvement in South-East Asia will further diversify trade links. These will be moved further from traditional patterns by, for example, the growth of air cargo movements: by 1975 it has been estimated that the tonnage involved will be around 75 000 tons or ten times the 1965 level.[7] Tourism, too, will become an important element of international trade. Luzon, and Manila in particular, now is too entrenched to lose its position as the focal point of international trade, especially as it can expect to handle virtually all the growth in air cargo. Nevertheless, existing trends towards the by-passing of Manila, particularly where bulk cargo exports are concerned, doubtless will accelerate and it may be that, for example, demand from the hinterland of Cebu may warrant direct delivery of imports. Certainly Cebu might be able independently to generate trade in both directions with Indonesia thus restoring in a small way the pre-Spanish trading pattern which at present is dominated by cigarette smuggling.

Too much should not be read into the above, however, for in the 1970s U.S.-Philippine trade will continue as a major aspect of international trade whatever the outcome of the ending of the Laurel-Langley Agreement and the current probability of further significant expansion of copper and banana exports to Japan. By the end of the decade this might no longer be the case but, as is demonstrated below, this is by no means certain.

U.S.-Philippine Trade. Unless present circumstances are altered significantly it would appear that the ending of the Laurel-Langley Agreement will severely curtail Philippine exports to the United States, and sugar in particular. The resultant reduction in foreign exchange earnings would disrupt current plans for economic development. Sugar, for example, is the largest single source of government revenue. There is increasing agitation for an extension of the Agreement (and more in the case of the sugar producers who want their quota increased). It is claimed that, if this were granted, it would give the Philippines adequate time to diversify and expand trade with other parts of the world, and to put trade with the United States on a firmer footing. On the other hand, there is a feeling in the United States that the trade benefits accorded to them by the Philippines in large measure had been nullified by the government's various fiscal measures introduced over the years to bolster the country's weak trading position. Furthermore, it is considered that until producers are exposed to the full realities of free trade, there will be inadequate stimulus for them to improve their competitive position. Once again, sugar is the outstanding example.

Future developments await the results of efforts by UNCTAD to facilitate access by developing nations to the markets of developed nations. Such an extension of non-reciprocal preferential tariffs would undoubtedly ease the Philippines' trading position. If this does not come about a compromise solution will be needed to break the present impasse over the Agreement. Unfortunately, the Philippines is in the weaker position and will need to rely heavily on past goodwill to secure a satisfactory result. In all probability, a short term extension of the Agreement, possibly restricted to a limited number of commodities, will result. Its period of extension will depend less on economic considerations than on the political situation in the area. An alternative arrangement would be a stepping up of economic aid designed to improve the competitive position of those industries worst hit by the ending of the Agreement. Sugar, of course, is outstanding in this respect. Any long term extension of the Agreement would appear to be out of the question for this would seriously undercut the United States' avowed intention to phase out discriminatory trade preferences currently extended to various of the world's developing nations.

No matter how strong past goodwill, therefore, it is apparent that in the 1980s and perhaps sooner, the Philippines will have to stand on its own feet in international trade. In economic terms the country, however well indirectly cushioned by the United States, may suffer initially, but the long term should see the emergence of viable export industries of great benefit to overall economic development. Sugar may be included in this category but probably on a smaller scale than at present. In social terms, the ending of the Agreement could well symbolise the final cutting of material ties binding the Republic to its former master and, by lessening the at present overwhelming Americanisation of its culture, lead more rapidly to the creation of a truly Philippine society.

Turning to patterns of domestic commerce, their future hinges primarily upon developments in transport where wholesale trade is concerned and upon rising living standards where retail trade is involved. Of the case studies presented, it is suggested that change will be more evident where commodity marketing is concerned in view of impending progress in the field of transport. Also, where retail trade is concerned, a relatively slow upward movement in living standards is predicted which will do little to accelerate current gradual trends towards sophistication of consumer demand.

In the field of commodity marketing there is a particular need for more efficient methods of moving bulk commodities, whether by land or sea, particularly so that food deficit areas of the country can be readily serviced as and when the need arises, and so that export prices of raw and semi-processed materials be minimised. Assuming that political and other non-economic considerations do not jeopardise the action of market forces, current and proposed improvements could go far to achieving these objectives.

Better transport implies also better market intelligence which is a function both of a branch of transport (telecommunications), which operates with reasonable and increasing efficiency, and of commodity marketing expertise itself. The latter operates with tolerable but decreasing efficiency as inroads are made into the influence of Chinese middlemen by government and quasi-government organisations and schemes which are the product of economic nationalism rather than economic efficiency.

The emergence of new marketing organisations, notably in the field of rice and maize, is in theory designed to simplify distribution channels thus both eliminating the Chinese middleman and unnecessary waste at the same time. But, in practice, the new organisations cause the Chinese businessman's profits to be replaced by graft and corruption by the politicians who direct and the civil servants who operate the marketing arrangements. Furthermore, operating inefficiency causes waste to be an even greater problem than in the past.

The 1970s, therefore, can be expected to be a period of transition. New and old methods of commodity marketing will exist side by side and it will not be until the former have proved their worth that the era of the Chinese middleman will be at an end. If past experience is any guide he still will be firmly entrenched at the end of the decade. Therefore, it is in the physical structures associated with commodity marketing that the greatest changes will occur in the 1970s. In this sphere transport facilities will take pride of place but improvements and innovations in processing, packaging and storage will be significant also as the predicted situation in Hilongos reveals.

Commodity Marketing. The present pattern of commodity marketing in Hilongos has shown little variation over the years nor are its bases likely to undergo any radical alteration in the foreseeable future despite government intervention in the form of the Rice and Corn Board. Nevertheless, it will not be unaffected by the nation's overall economic development. The result will be improvements in its efficiency and minor modifications to the rural scene.

Improvements in efficiency will be highlighted by road development works. The upgrading of existing feeder roads and the provision of additional kilometrage will improve links between the producer and the middleman, while the latter, for example, will be better placed to sell on the Cebu market thanks to the restoration of wharfage facilities at Hilongos and the increased frequency of air services. Processing methods can be expected to undergo a gradual change as more advanced techniques come within the budget of the middlemen. Outstanding in this respect would be the introduction of refrigeration in fish processing and, if the proposal comes to fruition, the establishment of a sugar *central*. Finally, the marketing of industrial crops will feel the effects of the drive for improved export quality in, for example, improved abacá grading standards.

Greater efficiency in commodity marketing could banish current features of the rural scene such as the sun drying of copra. Most prominent of the new features will be those associated with road transport, such as further petrol stations incorporating servicing facilities and a high level bridge across the Salag River, but new processing and storage facilities will make a contribution. Nor should it be forgotten that the marketing pattern will reflect changes in the economic importance of the various commodities; in this respect abacá could become a less conspicuous element.

Retail trade patterns in the Philippines reflect closely the general tenor of life and livelihood at the local level. In the normal course of events established patterns are slow to change especially in rural areas, or indeed even in many provincial centres. It needs the stimulus of an economic development scheme to cause sudden change in the pattern of retail trade. A community's proximity to such a development node, whether it be a newly exploited major mineral deposit or merely the construction of a new sugar *central*, has the effect of exposing it to an injection of substantial spending power, particularly by the often more sophisticated newcomers. However, a development node's impact is strictly localised; the case of Hilongos outlined below is thus typical of the majority of Philippine communities.

Retail Trade. The pattern of retail trade in the Philippines, as exemplified by the situation in Hilongos, can trace its origins to the pre-Spanish era. Subsequent modifications have evolved gradually and, although the pace of change undoubtedly has quickened, its pattern in the future will differ from that at present in matters of detail rather than of substance. Such changes that do occur primarily will reflect slowly rising productivity and hence living standards, although other factors such as new techniques and increased consumer mobility will play their part. It is not expected that in the coming decade disposable incomes in Hilongos will have reached a level whereby current luxuries become necessities, or specialist stores will have superseded the market, or that the supermarket will have replaced the general store.

Better storage and display facilities, notably refrigeration and more substantial market structures, are the most probable developments which could alter significantly the form of Hilongos retail trade in the 1970s. But the most flexible aspect of retail trade will be the nature of the goods offered for sale. Each year sees an improvement of the range, both in terms of new products and product quality, and this trend can be expected to continue in the future. Its pace primarily will reflect that of the country's overall economic development. For example, improved transport facilities may make available durian and mangosteen, fruits normally which are only sold at their point of origin in Mindanao, or in Cebu. However, it should not be forgotten that the Filipino's propensity for consumer goods of a sophistication more advanced than the nation's development warrants (colour television in Manila being the classic example) may well, even in Hilongos, cause to be offered for sale items such as imported biscuits and chocolate which appear hardly appropriate to a far from prosperous agricultural community in the tropics. With increased consumer mobility, demand for such items can be expected to grow at an above average rate although available income will restrict the quantities involved.

Greater sophistication of demand can be expected to lead to some increased importance for the specialist stores and a consequent decline in the general stores. Chinese control of the latter will wane also. A widening of the hinterland as a result of improved transport facilities will on balance increase turnover; peasants in the hinterland will shop more often in the town, more than offsetting the fact that its more substanial citizens will be induced to shop further afield on occasions. Greater mobility of the suburban dwellers could lead to some decline in importance of *sari sari* stores but this is not likely in such a small centre as Hilongos; rather, some could mature into general stores and form the nuclei of subsidiary shopping centres as the town grows in size.

B. Social Geography

Change in Philippine social geography takes place slowly. As already noted in the case of retail trade, patterns at the micro, or 'grass roots', level are conservative by nature. At the macro level, on the other hand, policy making and implementation in social matters frequently is handicapped by vested interests while government expenditure on social services, at 10% of total income, functions on a very restricted basis. The most volatile areas of social change thus are in 'pioneer' areas of the country where a lack of indigenous behaviour patterns coupled with the often diverse origin of the pioneers makes for a relatively high degree of flexibility. It is still to be hoped that a changed attitude to birth control will have an impact on demographic patterns. On paper there are ambitious schemes for family planning clinics (2000 by 1976 to serve 3 million women and thus cut the population growth rate to 2·3% per annum) but while those established in 1970 and 1971 have had some success the 1976 target appears overly optimistic to say the least.

The pioneer areas can be either rural or urban in nature Indeed of those aspects of social change of particular relevance to Philippine geography it is those which are urban oriented which are of most interest. This stems from the fact that the overriding interest of the social geographer in rural patterns of the 1970s concerns land reform rather than land settlement, a field for which prospects for progress in the foreseeable future are not encouraging. In the urban areas social change is very much a product of demographic trends—urban migration, the increasing youthfulness of the population, etc. From a geographical standpoint their implications are highlighted by the suburbanisation of urban society and the Filipinisation of upper and middle management. As will be demonstrated below these two developments could have a profound impact on the social geography of the Philippines in the 1970s.

In the political arena volatile elements abound. They are characterised, however, by their impermanence and thus limited impact upon Philippine social geography. New political leaders, new provincial boundaries, and new taxes in general leave undisturbed the Filipino's way of life. These developments furthermore divert attention from the need for fundamental political reform. The position is particularly discouraging for progress in President Marcos's unprecedented second term in office has been inauspicious, thus refuting the argument that continuity of government would lead to a more enlightened attitude on the part of the nation's leaders. Likewise, the emergence of political consciousness on the part of the university students has had limited impact despite, or perhaps because of, its militancy.

Any changes initiated in the political arena first will have to overcome the activities of pressure groups with vested interests. Only then, perhaps, could positive action be stimulated, for example, to accelerate advancement of the Moros and other minority groups, and the betterment of the workers of the sugar lands. In the interim, selected sectors of the community such as the large landowners and selected regions such as the Ilocos will benefit from their superior political expertise. For the immediate future it is international not local relations which appear likely to cause a reappraisal of attitudes leading to changes in the way of life in certain parts of the country. In this respect the Sabah issue is the most pertinent but, as will be demonstrated, the implications of the dispute with Malaysia also involve much broader issues.

The Rural Scene

Land Reform. Filipinisation and other aspects of social progress such as improvements to minimum wage legislation, and perhaps a greater government interest in health insurance and pensions, can be expected to improve living conditions for urban dwellers. On the other hand, prospects for rural dwellers are far from promising, with the population growth rate remaining out of control. Physical overcrowding and resultant crime and environment pollution may not afflict the rural inhabitant but instead probably more than one in three will become under-employed and many families will have to intensify the labour sharing arrangements and employment creating tasks. The stage might even be reached whereby, *reducto ad absurdum*, Javanese might be employed to demonstrate how in neighbouring Indonesia the activity of running even faster while standing still has been brought to a fine art.[8] More realistically, the 1970s will see greater control of the environment as PACD and other organisations pursue their activities; greater sophistication in farming attitudes as the extension services are improved; greater income for some, particularly those utilising HYV rice; greater security of tenure for some, given an absence of obstruction on the part of the landlord, and greater awareness of non-parochial matters as increasing numbers of friends and relatives move to the towns and cities. But, in almost every case, material and non-material gains made will have to be shared amongst more people. As a general rule by 1980 each *poblacion, barrio,* and *sitio* will possess at least four persons for every three to be found there in 1970, and even more where below average numbers move to the cities or to the rural resettlement areas.

The other root cause of the impoverishment of rural society—land tenure—seems unlikely to be much nearer solution in the foreseeable future either. This judgement is based on the consideration that measures initiated in the 1960s like preceding ones have achieved, at best, only marginal success. The Philippine agrarian system is undergoing a transition from a static subsistence structure to a technically progressive yet small scale commercial structure. Therefore, as Ruttan[9] concludes, there is not necessarily a clear relationship between productivity, tenure and farm size. Contrast this statement with the conclusion of Schul[10] that sugar cane yields are closely related to *hacienda* size; in this case of course one is dealing with the largest scale and most commercialised sector of Philippine agriculture. For most Philippine agricultural areas potential productivity gains are likely to be achieved only in association with a broad based agrarian policy which provides for adequate credit, technical information diffusion, etc. In other words if the Agricultural Land Reform Code is to be successful it must be implemented on all fronts, a task probably beyond even the most well intentioned of its administrators.

The magnitude of the implications of successful land reform in the Philippines can be judged from the work of Scott and Kuhonta.[11] They concluded that reorganisation of farms in

Laguna Province to ensure the universal provision of units of minimum economic size would involve:

(a) a 17% reduction in the number of lowland rice farm units
(b) a 35% reduction in the number of upland rice farm units
(c) an 18% reduction in the number of coconut farm units
(d) a 37% reduction in the number of sugarcane farm units.

The authors further expressed little optimism that the landholders involved (the farm units of Miguel and Juan, discussed in the previous chapter, would be amongst the early casualties if this policy were to be implemented) would be willing, or even able, to secure alternative land meeting minimum size requirements. As a result, there would be a migration of some 5000 families into nearby cities for which existing retraining facilities and employment opportunities would be totally inadequate.

Basically, the land tenure problem has been too complex, too deeply rooted and too universal for solution by the leaders of the Philippines since independence. It needs to be stated also that, despite the numerous investigations and analyses of the problem, by geographers and non-geographers alike, only an incomplete and possibly misleading picture of the situation as it affects the man on the land has emerged.[1,2] Even once full appreciation of the situation is achieved, land reform can only be achieved by the Filipinos and such an achievement can only be attained when Filipino society in general has reached maturity. In the 1970s as in the 1960s, any attempt at a solution will be bedevilled by, to name but a few causes, self-seeking attitudes on the part of key individuals, the exploitation of tenants' ignorance, and a lack of capital to foster the initiative of newly created smallholders or, alternatively, to overcome the persistence of conservative attitudes which dissipate the benefits of former tenants' newly gained independence.

Thus in terms of social geography little basic change is envisaged in the rural scene. Such a persistence of the *status quo*, of course, excludes the impact of any unpredictable developments such as a major spread of Huk type activity or, to quote a hypothetical example, a major policy decision to expel U.S. Peace Corps workers in retaliation for an unsatisfactory conclusion to negotiations concerning the Laurel-Langley

Agreement. In fact, as already indicated when discussing retail trade, increasing agricultural efficiency means that the rural way of life gradually will become more sophisticated as regards consumer expenditure. Likewise, communal access to amenities such as piped water, electricity and banking facilities will become more commonplace. In this sphere, too, the expected peripheral developments in land reform will contribute to both rising living standards and also a more civilised way of life.

It is of some significance that those areas in which the most far reaching changes to the rural way of life may occur, such as Central Mindanao and the Bondoc and Bicol Peninsulas, are those with relatively minor land tenure problems. As was the case with retail trade, the initiation of substantial change in the rural way of life depends on proximity to development nodes. One of the prerequisites for the establishment of such nodes is an absence of land tenure problems, which ensures both a favourable climate for new investment and the presence of a less conservative and more energetic work force.

Where development nodes are located in regions with serious land tenure problems such as Negros Occidental they are usually motivated by vested interests who are able to exploit the semi-feudal situation. Consequently, little benefit accrues to either the livelihood or the way of life of the rural inhabitants. Only occasionally are development nodes the products of less selfish attitudes; then, land reform forms an integral part of development as the Kaisahan, Negros Occidental, project illustrates.

The Kaisahan settlement has an area of 693 hectares and lies 145 kilometres south of Bacolod City (Fig. 46). The Roman Catholic Bishop of Bacolod initiated the scheme, whose primary objective is to ensure that every farmer owns the land he tills and at the same time has access to technical advice, credit facilities and marketing facilities, as well as schools, health centres and associated social infrastructure. Funds for the project, totalling several thousand pounds, include donations from private individuals, notably the Bishop of Bacolod, and the Philippine Sugar Institute. They are administered by the Kaisahan Settlement Foundation whose secretary has been the key to the community's inception. It was he who effected the release of the settlement area from the Bureau of Forestry and who arranged for the land to be tilled. Later it will be subdivided and awarded to the settlers.

Forest clearance was facilitated by two tractors acquired by the Foundation. These were hired out at cost to the settlers to be paid with the proceeds of their products after harvest. The

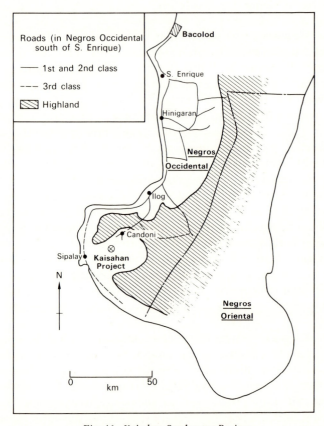

Fig. 46. Kaisahan Settlement Project

Foundation, through the cooperative which it is organising, will handle the marketing of the produce. Other Foundation funds are used to buy seeds, fertilisers, *carabao*, etc., to establish a small cattle ranch of sixty hectares and seventy head of cattle and, with the use of a bulldozer lent by the provincial governor, to construct two kilometres of road linking the settlement to the local highway. The road, however, is passable only during the summer when the streams along the route are dry.

The project has improved greatly the farm practices and attitudes of the settlers. Agronomists hold farmers' classes at least twice a month and, as a result, the settlers now are more receptive and even eager to learn modern farm practices. Furthermore, a farmers' association has been organised in preparation for the establishment of a co-operative which later on will handle all the affairs of the settlement. The scheme, though in its infancy, has already done a great deal towards improving the lot of the people in the settlement. More important, it has given the inhabitants an awareness that a better, brighter and happier future lies within their grasp. The same cannot be said of the majority of the rural inhabitants of Negros Occidental.

Transmigration. Such are the population pressures in the Philippines that there is a real danger that current desirable land holding sizes (average four hectares) allocated to settlers in Mindanao and other pioneer areas will be reduced to accommodate more migrants.[13] The government may be able to resist such a move but it will be able to do little to prevent established settlers from subdividing their holding amongst their next of kin. In one way or another therefore rural Mindanao should continue throughout the 1970s to ease the pressure stemming from the nation's abnormally high population growth rate.

It is however in the towns of Mindanao that greater prospects lie in relieving population pressure. Given that the economic development characteristics of Iligan can be emulated on a smaller scale elsewhere, the resultant employment opportunities could well enable the immigration rate to persist though doubtless a different type of migrant would be attracted, perhaps from overcrowded Manila, leading to the creation of a distinctive element in Mindanao society. The timber complex at Bislig Bay on the island's east coast is perhaps a prototype; such development nodes may not rival Iligan in size but such is the breadth and geographical spread of the island's potential resources that in total they could make a significant contribution to its social and economic geography.

The Urban Scene

Suburbanisation. By 1980 approximately 45% of the nation's population should be urban dwellers. This compares with some 20% in 1970. For the majority, their dwelling place will leave much to be desired but suburban living will become a reality for ever increasing numbers, a trend most evident in the metropolis of the Philippines. In Manila the greater relative concentration of middle class wealth will be an important factor in suburbanisation, but the impact of centrifugal factors currently at work in the metropolis will be accentuated. This state of affairs appears inevitable for the influx of population will continue as a result both of over-population in rural areas and the employment opportunities created by the anticipated growth in secondary and tertiary employment in the metropolis. Added to this will be the effect of continued improvements in transport and local authority intervention, although probably at a rate inadequate to keep pace with population increase. Furthermore, increased sophistication will also weaken the inertia of the Filipino inhabitants (though perhaps not the Chinese so much) of the inner metropolis.

How far and how fast these developments will influence the metropolitan pattern of suburbanisation remains uncertain. It is safe to assume, however, that the bulk of new settlement will be characterised by an absence of planning leading to an accentuation of the urban sprawl already so evident in the metropolitan area.[14] National and local attitudes towards the physical planning of Philippine cities leave much to be desired for they are at the mercy of political pressures. As a result, most existing legislation is either inadequate to deal with the situation or else is frustrated through lack of funds or executive power. Even if a local authority is able to take action within its boundaries the success of its measures is usually jeopardised by a lack of co-operation elsewhere. This can be due either to the difficulties outlined above in respect of existing legislation or, more crudely, by political antagonism. Thus, for example, the City of Manila was successful in evicting a large body of squatters (11 000 persons) from Intramuros, despite the initial refusal of neighbouring authorities to allow them to pass out of the City. Yet it received no cooperation from the authorities in Bulacan Province, twenty-five kilometres away, to which by agreement the squatters were transferred. The result was that the people involved were left to fend for themselves thus creating a new slum area. Indeed, about half drifted back into Manila subsequently for the lack of local employment opportunities at the new site committed them to expensive and time consuming commuting.

Whatever the success of physical planning measures some decline in importance of two of Manila's four original foci appears likely. Firstly, the Central Business District will suffer from pressure on existing available land, inadequate transport links with the outer suburbs, and competition from business development in Makati. Secondly, Malacañang Palace will be come less of a focal point as the government, and possibly the diplomatic institutions which surround it, move to Quezon City. In neither case is it expected that the consequence will be catastrophic. Not only will the processes be slow but they will be offset in part by, in the case of the Central Business District, a need to service the growing demands of the Port, and an opportunity to expand southward into the areas vacated by government institutions. As for Malacañang, the persistence of a highly centralised form of national government will ensure that the magnetism of the Presidential Palace is little diminished while the increased sophistication and living standards of the Manila population should steadily enhance the district's role and that of the Central Business District as the cultural and entertainment centre of the metropolis.

The impact of these economic and demographic trends could be such that by 2000 A.D., Manila's boundaries could reach out northwards as far as Malolos and Angat, eastwards to Antipolo, and south to Biñan (Fig. 47). In other words, the metropolis would encompass an area of some 1000 square kilometres compared with 200 square kilometres at present. And this does not take into account ribbon development, already a major feature of the periphery of the metropolis. This, if not controlled, could reach out to such distant centres as Calumpit (forty-five kilometres north from the centre of Manila) and Los Baños (fifty-five kilometres south). At the other end of the scale development nodes may have the power to create significant suburban areas in the foreseeable future. At Bislig, Mindanao, a formerly sleepy pioneer town of less than 5000 inhabitants now can look forward to a population in the

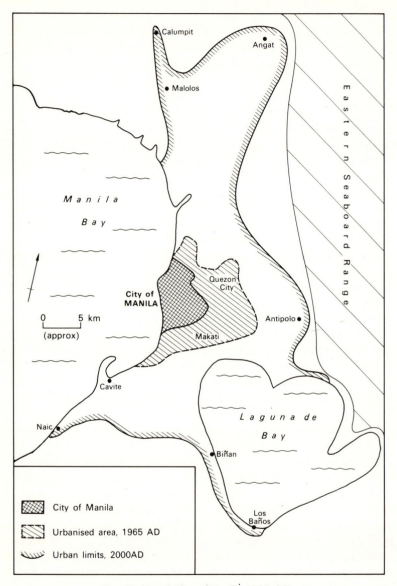

Fig. 47. Sketch Map of Manila, 2000 A.D.

immediate vicinity exceeding 75 000 once the entire timber processing complex is operational. More important, while the bulk of the labour force consists of unskilled or semi-skilled personnel, the community will number amongst its members a sizeable number of technical and managerial staff, together with the more normal élite of such a small community—lawyers, doctors, etc., which will generate a need for suburban development. This move will be reinforced by the fact that, being originally only a pioneer settlement, the town centre is not an attractive place of residence, as is the case with long established settlements.

It is to be hoped that modest but well executed real estate developments will fill the new suburban areas. Some such developments doubtless will be present, for the country possesses the appropriate architectural and allied talents. Unfortunately, it is all too probable that, at least in the 1970s, the imperfections of Philippine society in general will permit both ill-conceived and speculative building together with the presence of large squatter areas. The latter may receive the greater condemnation by the public at large but it is suggested that the former deserve the greatest criticism. Dwellers in squatter areas usually endure conditions little or no worse than in the village whence they came (the only basic difference between a slum in Tondo and Hilongos, for example, is one of areal extent). On the other hand, to provide the newly emergent middle classes with sub-standard dwelling places is not only setting an undesirable precedent for the nation's urban scene but also is creating an environment prejudicial to the establishment of a vital section of Philippine society and hence will gravely hinder overall social progress in the country.

Filipinisation. The 1970s could well see major advances in this field even if American business interests were to remain in strength through an extension of the Laurel-Langley Agreement. That Filipinos are more than competent to serve in middle and upper management was proven in the 1960s; in the 1970s an excessively youthful and educationally conscious nation such as the Philippines will contain far greater numbers of qualified personnel than at present. Failure to utilise to the full this labour resource not only will create serious employment problems but also will hinder the evolution of that most vital element of democratic society—the middle class.

All the indications are that employers will require additional managerial staff to cope with the expansion of business expected in the 1970s whatever the overall rate of economic growth. Whether demand will match supply is doubtful, thus accentuating present employment difficulties and perhaps exacerbating student unrest, but the 1970s should see· many educated Filipinos in stable employment and hence acquiring a comfortable suburban abode, and also a car, a T.V. set, or similar material assets. Whether they will make full use of their opportunities to create a truly Philippine segment of society rather than a psuedo-American one remains to be seen.

Whatever form middle class society takes it can be expected to alter greatly the urban, or rather suburban, scene. Such changes are evident already in Manila where housing estates, supermarkets, motorised mobility, etc., are their manifestations. Inter-regional centres such as Cebu and Iloilo can be expected to be in the van of such developments outside the metropolis. Pace setters in modern suburban living may be found also, though on a comparatively small scale, in recently established centres of private enterprise. Thus, for example, industry at Iligan City in the Misamis Coastal Belt, although possessing a much lower management to worker ratio than, say, Makati, can be expected rapidly to create sufficient middle management employment opportunities to create a demand for suburbanisation.

The suburban scene thus will show a trend away from the two extremes of Beverly Hills (Cebu)-type speculative estate development and the disorganised sprawl of traditional style urban housing towards hectare after hectare of unpretentious but, it is to be hoped, harmonious homes. On this score there is a need for sound planning both by local authorities and estate developers to benefit from experience elsewhere in the world. Furthermore, it is a field in which the urban geographer can make a valuable contribution.

Aside from the complications for urban geography, it is to be hoped that the steady infiltration of Filipinos into middle and upper management ultimately will lead to a forceful and constructive intervention into those spheres of society of which public service is the extreme example, where at present ability is not necessarily a prerequisite to the attainment of decision making status. Such intervention could occur both at the top, thus providing the positive leadership the country has lacked for so long, and at the community level, to provide greater finance, greater drive and initiative, and greater areal coverage than the present sprinkling of Rotary clubs, community minded society matrons, charity organisations, etc., are able to contribute. Such a development, on a nationwide basis, probably will have to await at least the end of the decade; it will only be then that Filipinisation can be considered a reality and that a major realignment of the social geography of the Philippines will have been effected.

The Political Scene

The 1970s undoubtedly will see a further weakening of American influence in the Philippines. Aside from increasing Japanese ties it seems likely that the Communist world will be accepted into the circle of Philippine international contacts. How far such contact will proceed is problematical but undoubtedly a significant leftward movement in Philippine international attitudes can be expected. This, however, should still leave the government of the day comfortably right of centre. The 1971 Constitutional Convention could herald a significant shift also, especially if it can provide concrete recognition of a desire for change widespread in the country at present. Significantly also the consensus of the very varied demands for change focuses on modification of the current American based electoral system.

Whatever shifts occur in political opinion in the 1970s one issue will remain unresolved, if for no other reason than that it is in the interest of the parties concerned that it should not be resolved. This is the Sabah issue.

The Sabah Issue. This feature of Philippine political geography appears unlikely to arrive at a permanent solution in the foreseeable future. Given stable internal conditions in the Philippines it could well prove conspicious by its absence[15] but it is only too probable that ultimately it will be raised again. The resultant furore probably will achieve nothing positive and will disrupt once again efforts to establish law and order in

South-Western Mindanao and set back regional unity in South-East Asia. In this respect any short term political advantages gained by raising the issue will be offset respectively by the damage done to the social and economic development of a region of the Philippines which has received too little of such development in the past and, on a much broader canvas discredit the international image of the country at a time when it will be actively seeking new trading partners.

From the point of view of Philippine regional geography, the continued uncertainty, at least in Filipino eyes, over the status of Sabah, postpones the inevitable reorientation of the way of life and livelihood of many of the people of South-Western Mindanao. In view of the damage being done to the national economy by smuggling, it is probable that, through some face saving gesture, the provisions of the anti-smuggling treaty will be implemented before very long. At that time there will ensue a marked acceleration in the process of curbing the nomadic and adventuresome instincts of the Tausog and Samal Laut. It is debatable whether their welfare will benefit but the region's pattern of occupance will certainly be modified with emphasis being laid on its tourist potential. Nationally, the boost given to the economy by control of smuggling will have important short term repercussions but these could be overshadowed, in the long run, by the reduction in the power of the criminal elements which currently exert undue influence in Filipino affairs.

Admission by the Philippines that, whatever the merits of the Sultan of Sulu's claim, the Malaysianisation of Sabah is now a *fait accompli* appears impracticable at this stage, for national pride, always exceptionally sensitive, is at stake. An end to the issue will have to await the full maturing of Philippine society when, by putting regional ahead of national interests, the concepts of ASEAN will have been vindicated. In the foreseeable future, however, notwithstanding the potential for friction between Malaysia and the Philippines, ASEAN could well prove to be the prime vehicle whereby traditional Philippine foreign policies are modified. Aided by, and at the same time causing, the gradual weakening of existing economic and cultural links with the United States, closer contact by the Philippines with its South-East Asian neighbours ultimately should lead to the emergence of a society attuned, though by no means exclusively, to the mores and needs of the Orient.

References

1. T.G. McGee, *op. cit.*, pp. 15-28 and 171-5.
2. H. Stockwin, 'The Widening Chasm', *Far Eastern Economic Review*, Sept. 19 1968, p. 594.
3. F.H. Golay and M.E. Goodstein, *Rice and People in 1990, Philippine Rice Needs to 1990: Output + Input Requirements*, USAID, Manila, 1967. pp. 10-11
4. *Ibid.*
5. M.L. Gupta, 'Manpower Plan for the Philippines: An Approach (F.Y. 1971-1974)', *Philippine Economy Bulletin*, Vol. 7 No. 6, 1969, p. 24.
6. Anon: *Far East Air Traffic: The Challenge of the Future*, Boeing Commercial Airplane Division, Renton, Washington, 1967, p. 12.
7. *Ibid*, p. 44.
8. C. Geertz, *Agricultural Involution: The Processes of Ecological Change in Indonesia*, University of California Press, Berkeley and Los Angeles, 1970.
9. V.W. Ruttan, 'Tenure and Productivity of Philippine Rice Producing Farms', *Philippine Economic Journal*, Vol. 5, 1966, pp. 46-63.
10. N.W. Schul, 'Hacienda Magnitude and Philippine Sugar Cane Production', *Asian Studies*. Vol. 5, 1967, pp. 258-73.
11. J.T. Scott Jnr. and P.C. Kuhonta, 'Resources Required for Viable Farms in Laguna Province, Philippines', *Illinois Agricultural Economics*, Vol. 8 No. 2, July 1968, pp. 29-35.
12. N.W. Schul, 'Problems in Land Tenure as viewed from a study of the Visayan Sugar Industry', unpublished M.S. no date, p. 12.
13. P.A. Krinks, 'Peasant Colonisation in Mindanao', *Journal of Tropical Geography*, Vol. 30, June 1970, pp. 38-47. In addition, virtually uncontrolled squatting, often encouraged for mercenary and political ends by the people appointed to protect the resources of Mindanao, makes possible the accommodation of more migrants. Such migrants are marginal in every sense of the term. Their occupation of the land will in the long run benefit neither themselves nor the nation. The only beneficiaries will be the initial pioneers who now, thanks to exploitation of opportunities present in the early days of settlement, possess the capital and influence to prey upon later arrivals in a manner similar to that of landlords of Luzon and elsewhere from whose grasp the migrants had hoped to escape.
14. One valuable counter-measure to urban sprawl would be the more efficient utilisation of existing urban land. Redevelopment, of course, is dependent on better planning control but there are also large areas of idle land being held for speculative purposes. Given greater success in collecting existing property taxes than in the past, landowners might have an incentive to satisfy the ever increasing demand for urban land.
15. The scale map of the Philippines which is a conspicuous feature of the New Luneta in Manila affects a discreet compromise by outlining but a small part of Sabah though at the same time reserving ample space to display the entire territory if so desired.

10
Conclusion

Increasingly it is the Filipino himself who assumes the premier role in shaping his country's economic and social geography. The country's unique physical features still form a limiting frame of reference but the Filipino now has much freedom to operate within these limits and, at times, to challenge them. This arises both because of the Filipino's greater technological expertise and also because decision making no longer is concentrated in the hands of expatriates. In economic terms the latter is of particular significance, for developments such as the elimination of the Chinese from retail trade, the Filipinisation of industrial management and, at the highest level, the gaining of full self-government has firmly placed in the Filipino's hands the prerogative for decision making. In social terms the Filipino has much less independence of action: he remains very much a prisoner both of long imposed and basically exotic behaviour patterns, in particular those of Spanish origin, and also of more recent, but not necessarily superficial, cultural influences of trans-Pacific origin. It is suggested however, that the time is approaching when independence of decision making in social fields can be expected to be asserted. This particularly applies once the novelty of economic independence fades, or perhaps is replaced by disillusionment if the unsatisfactory economic and financial climate of the post-independence years persists.

Future attitudes to economic and social decision making will be governed by many factors but preeminent is the character of the Filipino himself. Patterns of economic and social geography in a newly independent developing nation such as the Philippines can be expected often to be the product of decisions, or indecisions, based at times not upon an informed or realistic response to the environment but more upon political, emotional, nationalistic and similar considerations. In the recent past, for example, the failure of land reform, the neglect of the Cagayan Valley, the Sabah issue, and industrial over-capacity and fragmentation stem from ill-conceived decision making. Against this, of course, must be set successes in decision making such as the development of industry in the Misamis Coastal Belt, coconut planting in Mindanao, HYV rice, and the growth of Makati. Unfortunately, such successes have not been sufficiently numerous or widespread to cause profound changes in the pattern of Philippine economic and social geography.

Past and present influences upon the Filipino's character often are extremely deep seated; religion and kinship ties are perhaps outstanding in this respect. As Sternberg[1] points out: 'No one questions the Filipino's patriotism, his devotion and loyalty to kin and place of origin, but his idea of 'kind' is still too centered on family and kinship, his idea of place too narrowly confined to the province or region of his birth.' The fact that such attitudes may override economic considerations contributes to, for example, the reluctance of many Tagalogs to settle permanently far from Manila; the self-centred attitude of the sugar barons of Negros Occidental; and the prevalence of under-employment in commerce, industry and public service.

Filipino behaviour patterns are also influenced by the inhibitions created by the matriarchal basis of Filipino society; the omniscient guidance provided by the Roman Catholic church; unfamiliarity with the outside world other than the United States; and, perhaps subconsciously, a feeling of inferiority related to the fact that independence, at least from the Americans, did not necessitate the aggressive, or indeed violent, nationalistic behaviour characteristic of many Afro-Asian nations.

The above facets of the Filipino character contribute in no small way to current patterns of life and livelihood. Perhaps most striking are the excessive commercialism of the Filipino way of life, the flamboyant and, at times, aggressive behaviour of some Filipinos, and, in a more pleasant vein, the people's friendliness, graciousness and personal smartness. Similarly, the Filipino character contributes to such diverse features as the proliferation of cheap and gimmicky non-essential items in retail trade, the excessive number of television channels in Manila, and the indifference shown to the conservation of the unique wild life of Palawan. But it is the Filipino's religious behaviour which epitomises the character of the nation—it has vigour, enthusiasm, expectations and faith but lacks realism, experience, and sophistication.

The Filipino thus lacks the trait possessed in no small degree by his former masters—the Spanish—which is an essential prerequisite of nationhood, that is, a sense of deep personal and national dignity. As Sternberg[2] puts it:

'There is need in the Philippines today for a vigorous and articulate nationalism, one which will subject the total society to expert scrutiny, determine its deficiencies and chart the course by which the people of this open society can most speedily and painlessly effect the transition from semi-feudalism to progressive nationhood. The need is for something more than a dilettantish commitment of the Filipino intellectual to his nation's growth and development. It calls for a wide-ranging public discussion of traditional values and behaviour patterns which retard or inhibit the development of institutions—social, political and economic—essential to a modern, industrialised democracy. Out of such public discussion should come the new values, the clearly defined directions and acceptable goals of the Filipino people.'

Currently, there is only one field in which this trait is well marked. Dancing and music are an essential part of daily life and to quote Kirkup[3]: 'Filipino dance seems to me a true expression of the native soul and the one indigenous cultural feature of a folklore nature that today reminds Filipinos of their national identity.' Otherwise, responsible leadership is the exception rather than the rule. Most important, only a few of the nation's decision makers can do little to combat the all too prevalent self-seeking, profiteering and corruption.

Nevertheless, there is cause for optimism. The Philippines possesses three valuable legacies from the American administration: a very high degree of political freedom, an articulate press, and a universal and usually egalitarian system of education. In the Philippines, therefore, although it is very far from the truth to say that all men are born equal or possess equal opportunities, one and all have adequate opportunity to make the most of their talents. This has been a very potent factor both in minimising Communist activity and, at the other end of the scale, preventing the Philippines from resembling a Latin American 'Banana Republic', as its Spanish and American origins might indicate.

Furthermore, in the 1970s, any Filipino, no matter how poor or uneducated, will have not only the opportunities but also the examples provided by current Filipinisation to encourage him to improve his lot. The imperfections of society still provide many obstacles to advancement but increasingly men of ability, as opposed to those brought to the fore by patronage, will come to occupy the decision making positions in society. By their education and expertise they will be able to mould the presently ineffectual middle class into a strong bridge between the extremes of Filipino society. The emergence of a strong middle class would be of immeasurable benefit to the Philippines for not only might it well mitigate many of the country's present problems, notably land tenure, crime and corruption, and public service inefficiency, but also the underlying stability of society so created would facilitate greatly overall economic and social development and greatly strengthen the nation's image and prestige overseas.

The purpose of this book goes beyond the traditional objectives of studies in economic and social geography. Concern with future patterns of Philippine economic and social geography requires that special attention be paid to the dynamic elements of the environment. Of these it is the Filipino himself who emerges as the key factor. In the years to come it is to be hoped that the maturing of his character will lead to a weakening of the presently strong influence of decision makers amongst whom selfish and narrow-minded attitudes predominate.

In the foreseeable future, developments affecting the pattern of economic and social geography in the Philippines will be dominated by those outlined in Chapter Nine. Many of these, if properly implemented, could significantly modify present patterns, but, even if this comes about, many imperfections will remain and many parts of the country will be little, if at all, affected. This is not to say that the establishment of developmental nodes be abandoned, but there is a need for their equitable distribution to take second place to developments having nationwide beneficial implications, particularly those which become readily apparent at the *barrio* level. It seems doubtful if such an approach will have prominence in the nation's economic and social planning in the 1970s for in this decade priority needs to be given to the Filipino coming to terms with his own character. Until he masters himself he cannot expect to master his surroundings.

References

1. D.T. Sternberg, 'The Philippines: Contour and Perspective', *Foreign Affairs*, Vol. 44, 1965-66, p. 506. Copyright of this article is held by the Council on Foreign Affairs, Inc. New York, N.Y.
2. *Ibid*.
3. J. Kirkup, *op. cit.*, 1968 p. 155.

Glossary

Alibangbang	pioneer, fire resistant, tree characteristic of second growth forest
Apitong	member of the dipterocarp family providing structural timber
Anee	shade tree employed by abacá producers
Bagoong	fish paste made from salted anchovy or shrimp
Balibol	rice dehusking implement
Bangos	milk fish
Barangay	pre-Spanish terminology for *barrio*
Barrio	administrative unit, usually a village
Basnig	small, bag-type fish net
Binayuyu	see *Alibangbang*
Bolo	multi-purpose cutting knife
Cacique	member of the former Philippine aristocracy
Cadang cadang	virus disease of the coconut
Calesa	horse-drawn carriage plying for hire in Manila
Camote	sweet potato
Canon	rent paid by leasehold tenant
Carabao	water buffalo
Carnapping	motor vehicle theft for subsequent ransom
Centavo	one hundredth of a *peso*
Central	centrifugal sugar mill
Cogon	tall, coarse, largely unpalatable, tropical grass
Comerciante	businessman
Cono	large rice mill
Daeng, tinapa, tuyo	varieties of preserved fish i.e., dried, smoked and salted, respectively
Dapdap	see *Anee*

Datu	pre-Spanish terminology for *teniente*
Encargado	manager of a *hacienda*
Encomienda	Crown land temporarily held by an individual
Fiesta	celebration of the local patron saint's day
Guijo	tree similar to *apitong*
Hacienda	large agricultural holding
Haciendero	owner of a *hacienda*
Hagotan	spindle employed to strip abacá
Halabas	coconut harvesting implement
Inquilinato	leasehold tenancy
Jeepney	small public service vehicle with a jeep-type chassis, usually ornately decorated
Impitan	main, or detention, areas of a fishpond
Ipil	source of cabinet timber
Kaingin agriculture	shifting cultivation
Kainginero	person who practises *kaingin* agriculture
Kanduli	freshwater fish found in Laguna Bay
Kasama	share tenancy
Katiwala	foreman or village elder
Kiskisan	small rice mill
Lauan	the softer members of the dipterocarp family, otherwise known as Philippine mahogany
Lawag	large, seine-type fish net
Lithao	fine toothed harrow
Loknit	abacá stripping implement
Mahjong	popular Chinese game resembling Canasta but played with ivory or plastic blocks
Molave	type of forest comprising semi-deciduous members of the legume family, and one such member which yields a variety of cabinet timber
Moro	adherent of Islam resident in Mindanao
Municipio	municipality, or administrative unit served by the *poblacion*
Mungo beans	green beans
Muscovado	loose brown sugar similar to, but coarser than, refined sugar
Narra	source of cabinet timber

Nipa	variety of palm inhabiting swampy brackish areas
Pabiayan	nursery for a fishpond
Pague pague	insect pest of abacá
Palay	the rice plant, or rice grain with husk
Patio	place where threshing of *palay* occurs
Patis	fish sauce
Peso	principal monetary unit
Plaza	park and open air meeting place for the community
Poblacion	administrative unit, usually a small town
Sampaguita	the national flower of the Philippines; it is small, sweet smelling and white in appearance
Sari sari store	small, general purpose shop
Sitio	discrete housing unit smaller than a barrio
Suki	the personal relationship existing between two businessmen well known to each other or between shopkeeper and regular customer
Tao	person, term often used when referring to farm worker or labourer
Tapahan	small copra drying kiln
Tapulau	variety of pine tree
Tartanilla	Visayan equivalent of *calesa*
Teniente	village headman or community leader
Tinapa	see *Daeng*
Topil	sugar planting implement
Tuba	alcoholic drink, a product of the coconut palm
Tuyo	see *Daeng*
Yakal	tree similar to *apitong*

Select Bibliography

J.N. Anderson, 'Buy and Sell and Economic Personalism: Foundation for Philippine Entrepreneurship', *Asian Survey*, Vol. 9, 1969, pp. 611-68.

Anon, *An Economic Profile: Philippines*, Sycip, Gorres, Velayo and Co., Makati, 1969.

A. Barrera, 'Classification and Utilisation of Some Philippine Soils', *The Journal of Tropical Geography*, Vol. 18, August 1964, pp. 17-29.

G.D. Berremean, *The Philippines: A Survey of Current Social, Economic and Political Conditions*, Cornell University, Southeast Asia Program, Ithaca, 1956.

Bureau of Census and Statistics, *Census of the Philippines* (1960), Manila, various dates. Also 1970 Census publications (as and when available).

Central Bank of the Philippines, Manila (annual report).

M.B. Concepcion (Ed.), *Philippine Population in the Seventies: Proceedings of Second Conference on Population, 27-29 November 1967*, Community Publications Inc., Manila, 1969.

H.C. Conklin, *Hanunoo Agriculture in the Philippines*, FAO, Rome, 1957.

O. Corpuz, *The Philippines*, Prentice-Hall, Englewood Cliffs, 1965.

A. Cutshall, *The Philippines: Nation of Islands*, D. Van Nostrand Co., Princeton, 1964.

L.B. Darrah and F.A. Tiongson, *Agricultural Marketing in the Philippines*, University of the Philippines College of Agriculture, Los Baños, 1969.

Economist Intelligence Unit, *The Philippines and Taiwan* (quarterly report).

S.C. Espiritu and C.L. Hunt (Eds.), *Social Foundations of Community Development: Readings on the Philippines*, R.M. Garcia Publishing House, Manila, 1964.

Fookien Times Press, Fookien Times Yearbook (annual report).

F.H. Golay, *The Philippines: Public and National Economic Development*, Cornell University Press, Ithaca, 1961.
F.H. Golay and M.E. Goodstein, *Rice and People in 1990. Philippine Rice Needs to 1990: Output and Input Requirements*, USAID, Manila, 1967.
P.G. Gowing and W.H. Scott (Eds.), *Acculturation in the Philippines: Essays in Changing Societies*, New Day Publishers, Quezon City, 1971.
G.M. Guthrie (Ed.), *Six Perspectives on the Philippines*, Bookmark Inc., Manila, 1971.
R.G. Hainsworth and R.T. Moyer, *Agricultural Geography of the Philippine Islands: A Graphic Summary*, U.S. Department of Agriculture, Washington D.C., 1945.
R.E. Huke, *Shadows on the Land: An Economic Geography of the Philippines*, Bookmark Inc., Manila, 1963.
R.E. Huke, *Bibliography of Philippine Geography, 1940-63: A Selected List*, Geography Publications at Dartmouth No. 1, Department of Geography, Dartmouth College, Hanover, 1964.
J. Kirkup, *Filipinescas: Travels through the Philippine Islands*, Phoenix House, London, 1968.
A. Kolb, *Die Philippinen*, Koehler, Leipzig, 1942.
T.G. McGee, *The Southeast Asian City*, G. Bell and Sons, London, 1967.
National Economic Council, *Philippine Economy Bulletin,* (bi-monthly publication).
K.J. Pelzer, *Pioneer Settlement in the Asiatic Tropics*, Special Publications No. 29, American Geographical Society, New York, 1948.
R.L. Pendleton, 'Land Utilisation and Agriculture of Mindanao, Philippine Islands', *Geographical Review*, Vol. 32, 1942, pp. 180-210.
J.L. Phelan, *The Hispanisation of the Philippines: Spanish Aims and Filipino Responses, 1565-1700*, University of Wisconsin Press, Madison, 1959.
Philippine Association, Weekly Economic Review (weekly publication).
Philippine Studies Program, University of Chicago: *Area Handbook of the Philippines, Human Relations Area Files*, New Haven, 1956.
J.H. Power and G.P. Sicat; *The Philippines: Industrialization and*

Trade Policies, Oxford University Press, London, 1971.

J.E. Spencer, *Land and People in the Philippines: Geographic Problems in Rural Economy,* University of California Press, Berkeley, 1952.

T. Tsutomo, 'Land Ownership and Land Reform Problems in the Philippines', *The Developing Economies,* Vol. 2, 1964, pp. 58-77.

E.L. Ullman, 'Trade Centers and Tributary Areas of the Philippines', *Geographical Review,* Vol. 50, 1960, pp. 203-18.

N. van Breemen *et alia,* 'The Ifugao Rice Terraces', in *Aspects of Rice Growing in Asia and the Americas,* Miscellaneous Papers 7, Landbouwhogesschool, Wageningen, 1970, pp.39-73.

F.L. Wernstedt and J.E. Spencer, *The Philippine Island World: A Physical, Cultural and Regional Geography,* University of California Press, Berkeley and Los Angeles, 1967.

Index

Abacá 71-73, 192, 197, 201, 210, 224, 258-61, 284-6, 329-30
Abacá Corporation of the Philippines (ABACORP) 260
Abacá and Other Fibres Development Board 260, 329
Aborigines 48
Advertising 245
Agricultural Credit Administration 297
Agricultural Extension 255, 322, 325, 347, 351
Agricultural Guarantee Loan Fund 297
Agriculture 42-80, 171-2, 174, 178, 187, 196-7, 199-200, 203, 210, 215-16, 219, 223-4, 227, 240, 251-61, 322-30
Agricultural Land Reform Code 155-6, 297, 300, 347
Air Cargo 339
American Administration 131, 181, 241-6, 295, 302, 361, 362
Ambuklao Dam 323
Angat Hydro-Electric Scheme 180
Animal Dispersal Act 258
Architecture 137-8, 355
Area of Land Mass 1
Asia Magazine 309
Association of Barrio Councils 25
Association of South East Asian Nations (ASEAN) 313, 339, 358
Australia 166, 339
Ayala Corporation 307

Balance of Payments 246, 250, 319
Bananas 216, 320
Bangkok 186, 225
Behaviour Patterns 125, 159-60, 237, 242-3, 292, 304, 345, 352, 361-2
Birth Control 142, 345

Bishop of Bacolod 349
Board of Investments 264, 336
British Government 312
British Isles 3
British North Borneo Company 311-12
Brunei, Sultan of 311-12
Bukidnons 231-2
Bureau of Animal Industry 258
Bureau of Forestry 349

California 202
Caltex Oil Company 136
Case Studies 6, 10, 283
Cash Crops 252, 256, 322
Cement Industry 178, 200, 330
Central Luzon Development Programme 297
Central Philippine University 130
Ceylon 332
Chinatown 18-19, 30, 306
Chinese 19, 24, 33, 125, 164, 210, 233, 241, 281, 285, 308, 310, 342, 352
Chromite 93, 178
Chrysler Motor Company 336
Coal 200
Coconut Research Institute 333
Coconuts 58-65, 188, 197 199-200, 201, 219, 224, 227, 253, 320, 324
Colonisation 1, 5, 163, 189, 242
Commerce 28, 33, 200, 220, 241, 277-91, 339-45
Commodity Marketing 27, 281, 283-6, 336, 341-3
Communism 165, 357, 363
Communist China 225
Community Development 20, 40-41, 357
Constitution, The 130, 135

Constitutional Convention 357
Copper 91, 191, 200
Copra Crushing 200, 265-8, 332-4
Corruption 133, 134-5, 342, 362
Crime 20, 133, 135, 346
Crop Combinations 45-46
Cultivated Land 42

Davies, Theo H. & Co. (Far East) Ltd. 309
Del Monte Pineapple Company 220
Department of Agricultural Extension 48
Department of Education 132
Development Bank 260, 268, 327
Development Nodes 331-2, 343, 349, 353, 364
Dialect 48, 128, 193, 199, 203
Disease (of Crops) 62
Dole Philippines Corporation 309
Drainage Patterns 97, 179, 188, 203, 206, 214
Drought 171-2, 199, 203, 218, 226

Economic Development 176, 196, 236, 292, 318, 338, 343
Economic Nationalism 164, 330
Economic Personalism (*Suki*) 160-2
Economic Plans 136
Education 3, 18, 36, 130, 131-3, 176, 243, 290, 363
Electricity 5, 13, 180, 320, 349
Employment 151-3, 183, 250-1, 261-4, 294-5, 307, 330, 346, 352-3, 356
Encomienda System 237-8
English Language 5, 128, 243-4
Entertainment 36, 275, 287, 289, 362
Europe 332
Expatriate Population 126, 308
Exports 250, 277-80, 318

Family Income 251
Far Eastern University 18
Farming Pattern 46-47
Farm Size 348, 351
Feeder Roads 78, 212, 217, 337, 342
Fertiliser 68, 253, 322, 325-7
Fiesta 162-3
Filipinisation 304, 308-10, 355-7, 363
Fire 21, 24
Firestone Tire and Rubber Company 270

Fiscal Policy 322, 340
Fishing 80-84, 189, 191, 197 205, 210-11, 219, 223-4, 228-9, 284-5
Fishponds 83, 178, 320
Flooding 75, 171-2, 212
Food Crops 48, 252, 256, 322
Food Production 318
Ford Motor Company 336
Foreign Investment 264, 330
Forestlands 84, 117-20, 191, 196, 203 222, 227, 323
Forestry 84-89, 171, 210, 224, 282
Forest Clearance 51, 87, 120, 255, 350
Four Year Plans 265, 272, 275, 318, 329, 336, 337
Friar Lands 238

General Motors 336
General Stores 281, 289, 344-5
Gold 91, 320
Goodrich Tire and Rubber Company 270
Goodyear Tire and Rubber Company 270
Government 135-7, 241, 243
Grasslands 120-1, 191, 232
Gross National Product (GNP) 3, 250, 318

Hawaii 202
Health 3, 244-5, 346
Hispanisation 214, 218
Hong Kong 225
House of Representatives 135, 312
Housing 25, 137
Hukbalahaps (Huks) 133, 179, 296-8, 348
Hydroelectricity 39, 176, 180, 219-220

Ifugao Rice Terraces 53, 174
Igorots 174, 176
Imports 250, 277-9, 318, 327
Import Controls 270, 277-9, 335
Indonesia 3, 125, 302, 339, 347
Industrialisation 38-41, 153, 187, 200, 220, 261, 305, 322, 332
Inflation 250
Inter Island Shipping 202, 224, 336-7
Inter Island Trade 28-29, 182, 200, 272-3
International Trade 182, 200, 272, 277, 281, 339

Index

Intertropical Front 104
Intramuros 16-17, 182
Investment 3, 266
Irian 3
Iron Ore 91, 191
Irrigation 50-51, 110, 173, 174, 180, 188, 239, 298, 324
Isneg, The 232-3

Japan 3, 43, 93, 142, 165, 277, 317, 357
Japanese Empire 207
Japanese Entrepreneurs 207, 211
Java 302
Javanese 347

Kaingin Agriculture 51-52, 117, 119-20, 175, 214, 227, 231, 323
Kaisahan Settlement Foundation 349-51
Kalimantan 3
Karst Landscape 102, 199, 226
Kinship Ties 151, 158, 232, 304, 361

Land Bank 156, 181
Land Capability 253
Land Ownership 23, 33, 153-4, 158, 236-8, 295
Land Reform 130, 133, 155-8, 246, 295-302, 345, 346-9
Land Reform Project Administration (Land Reform Authority) 156, 157, 297
Land Tenure 35, 50, 125, 153-8, 295-6, 300, 347-9
Land Use 43-46, 323
Language 126-8, 232
Laurel-Langley Agreement 279-80, 318, 322, 333, 339, 340-1, 349, 355
Law and Order 133
Legaspi, Miguel Lopez De 28, 234
Liberal Party 136
Literacy 3, 16, 126, 180, 290
Livestock 216, 256-8, 327-8, 350
Luneta 16-17

Magat Irrigation Scheme 173
Magellan, Ferdinand 28, 193
Maize 54-57, 171, 196, 199, 210, 215, 219, 224, 238-9, 320, 327
Makati 13, 307, 353
Malay Mongoloids 230

Malaysia 312-13
Malnutrition 142
Manila Galleons 181
Manufacturing 13, 183, 185-6, 200, 261-71, 283, 330-6
Markets 24, 281, 286-9
Mary Johnston Hospital 130
Matriarchial Society 158, 361
Meat 256-8, 327-8
Mechanisation 53-54, 66, 77, 82, 255, 322
Media 275
Mestization 126
Mexico 244
Middle Classes 19-20 304, 306, 356, 363
Migrants 20, 41, 66-67, 125, 128, 173, 201, 205, 217, 301, 351
Migration, to cities 149, 198, 302, 348
Migration, to pioneer areas 149-51, 198, 214, 219, 228, 302-3, 345
Migration, seasonal 147-9
Mining 89-93, 175, 191, 200, 211, 224, 228
Monsoon 179, 227
Moros 131, 134, 196, 212-13, 223, 225, 303, 346
Moro-Christian Unrest 215, 282
Motor Vehicle Assembly 268-71, 334-6
Motor Vehicle Density 180
Motor Vehicles 270, 276, 338

National Economic Council 251
National Land Authority 217
National Language 128
Nationalista Party 136
Natural Resources 5
Negritos 125, 230-33
Nett Domestic Product (NDP) 248-50
New Guinea 3
New Peoples Army (NPA) 296-8
Nickel 93, 211, 320
North Borneo 311-12
Northern Motors 270

Organised Labour 251

Pan-Philippine Highway 197, 275, 320, 334, 338
Parity Rights 164
Paper Making 329
Payne Aldrich Bill Tariff Law 279
Pesticides 255

Phil-Asia Trading Corporation 78
Philippine Sugar Institute 349
Philippine Coconut Administration (PHILCOA) 268
Philippine Rural Reconstruction Movement 297
Pilipino 128
Pineapple 216
Piracy 223
Planning 245, 352, 356, 364
Plantations 47
Planters Association 204
Plywood 264
Politics 1, 25, 130, 134, 135-7, 171, 176, 197, 205, 235-6, 242, 245, 287, 312-13, 342, 345-6, 352, 357-8, 362
Pollution 346
Population 138-53, 222, 294, 306, 347
Population, Age of 142-3
Population Distribution 145-6, 218, 227,
Population Increase 5, 125, 140-42, 145-6, 219, 245, 291-2, 345
Population Mobility 126, 147-51
Population Pressure 176, 347, 351
President Macapagal 295, 297, 312
President Magsaysay 296, 297
President Marcos 137, 270, 312, 313, 346
Presidential Act on Community Development (PACD) 292-3, 347
Procter and Gamble 266
Productivity 57, 252, 300, 322
Proto Malays 230-33

Racial Characteristics 5, 125, 233
Rainfall 75-77, 104-107, 109, 179, 188, 191, 196, 199, 203, 207, 214, 222, 226-7
Rainfall Reliability 107
Rat Infestation 54
Regional Analysis 6-7, 167-9
Religion 5, 36, 129-31, 140-42, 162-3, 176, 232, 234-5, 287, 361
Renault Motor Company 336
Republic Glass Corporation 309
Residential Patterns 22-26, 41, 291, 304-8, 352-5
Retail Trade 19, 27, 281, 286-91, 341, 343-5, 362
Ribbon Development 36, 305, 353
Rice 36, 48-54, 109, 171, 174, 178, 180, 188, 205, 215, 219, 227, 252-6, 281-2, 298-300, 320, 322, 324-5, 327, 347

Rice and Corn Board 310, 342
Rice and Maize Milling 23, 54, 57
Rizal, José 17, 187
Rubber 224
Rural Development 152-3

Sabah Issue 165, 222, 311-13, 357-8
Saigon 186
Salt 174, 178, 228
San Juanico Bridge 197
Sari Sari Stores 281, 291, 344-5
Savings 3
Silliman University 34, 133
Singapore 28
Slums 18, 40, 142, 304, 306, 346, 355
Smuggling 134-5, 165, 223, 225, 313, 340, 358
Social Communications Center 251
Social Intercourse 27, 36
Social Mobility 27
Social Structure 37-38, 128, 236, 304
Soils 51, 53, 60, 65, 72, 77, 112-17, 175, 191, 207, 214, 222, 227
Soil Conservation 121
Soil Erosion 55, 116, 117, 121-3, 175, 199, 218, 253, 322
South East Asia 1, 5, 125, 225, 300, 358
Spanish Influences 16, 28, 55, 126, 137, 165, 176, 181, 189, 196, 227, 234, 241, 295, 339
Spanish Language 128
Specialist Stores 290-91, 344
Squatters 20, 21, 34, 37, 323, 355
Sri Vijaya Empire 212, 233
State University of Mindanao 217
Steel Industry 40, 220
Student Activity 17-18, 346
Suburbanisation 183, 304-8, 352-5
Sugar 65-70, 178, 200, 203-4, 227, 279-80, 320, 323, 325, 340, 346
Sugar Exchange 204
Sugar Producers Marketing Association 327
Sukarno 136
Sulu, Sultan of 311-12, 358
Sumatra 302
Supreme Court 135
Swampland 99, 102, 179, 207, 213, 320
Synthetic Substitutes 71, 252, 260, 322, 329, 332

Tagalog Speakers 128

Taiwan 142, 318
Tax Evasion 134
Tectonic Activity 97, 187, 206, 230
Temperature 107-9, 196, 213-14, 222
Tenancy 133, 154
Textile Industry 175, 185-6, 189
Timber Processing Plant 87, 205, 219, 227, 351, 355
Timor 3
Tobacco 73-80, 171, 174, 238
Tourism 175, 176, 187, 222, 225, 274-5, 338-9
Toyota Motor Company 336
Transmigration 302-3, 351
Transport and Communications (All Modes) 3, 19, 172, 180, 192, 197, 198, 202, 212, 217, 224, 244, 272-6, 287, 304-5, 308, 333, 334, 336-9, 341-3, 351, 352
Tunku Abdul Rahman 136
Typhoons 25, 26, 60, 62, 72, 104, 190-194, 252-3, 268, 322

Unilever 266
United Nations 164, 312, 337
United Nations Conference on Trade and Development (UNCTAD) 318, 340
United States of America 138, 165, 241, 244, 250, 277, 279-80, 309, 332, 340-1, 357, 361
United States Congress 135
United States Peace Corps 348

United States Supreme Court 124
University of the East 133
University of the Philippines 13, 18, 132, 307, 328
University of San Carlos 30
University of Santo Tomas 18, 133
Urban Growth 13, 20, 30-31, 34, 177, 211
Urban Hierarchy 10, 34, 38
Urban Morphology 26, 36, 216, 240
Urban Settlement 175, 193, 204, 216-17

Vegetables 174, 215
Velasco, Geronimo 309
Vice President Lopez 136
Vietnam 1, 250
Virginia Tobacco Association 78
Volkswagen Motor Company 336
Vulcanism 97, 99, 187, 190-1, 213, 230

Wages 69, 260-1, 301, 346
Wallace Line 96
Watershed Conservation 89, 323
Wealth, Maldistribution of 251
Weeds 52
Wholesale Trade 19, 281

Yields, of Abacá 71
Yields, of Maize 57, 215, 252
Yields, of Rice 48, 52, 171, 215, 252
Yields, of Sugar Cane, 347